Social Network Structures and the Internet

Social Network Structures and the Internet

Collective Dynamics in Virtual Communities

Dongyoung Sohn

CAMBRIA
PRESS

Amherst, New York

Requests for permission should be directed to:
permissions@cambriapress.com, or mailed to:
Cambria Press
20 Northpointe Parkway, Suite 188
Amherst, NY 14228

Library of Congress Cataloging-in-Publication Data

Sohn, Dongyoung.
Social network structures and the Internet : collective dynamics in virtual communities / Dongyoung Sohn.
p. cm.
Includes bibliographical references and index.
ISBN 978-1-60497-536-9 (alk. paper)
1. Information technology—Social aspects. 2. Computer networks—Social aspects. 3. Internet—Social aspects. 4. Online social networks. 5. Communities. 6. Technology—Social aspects. 7. Electronic villages (Computer networks) I. Title.

HM851.S66 2008
303.48'33—dc22

2008019292

Table of Contents

List of Figures

List of Tables

Foreword

It is clear that the Internet is turning out to be something other than many believed just a short time ago. At the dawning of easy-to-use browsers, the thought was that this new medium would herald the advent of one-to-one communication and essentially do away with the notion of mass media. Some thought the old mass media such as television and newspapers would disappear at some point.

Of course, none of this has come to pass. Old media, history teaches, do not disappear when a new medium comes along. They change. Often, as has been the case with the Internet, the new medium first borrows methods and techniques of communication from the traditional media such as radio and telephony. Later, new means are invented to suit the special characteristics of the new medium. And, of course, these new methods invented for the new medium eventually find their way into the traditional media as these media change and adapt to the new media environment. In addition, the idea of mass media has not disappeared. If anything, the Internet has come to be viewed as a cross between earlier understandings of the issues involved in interpersonal and mass communication.

The idea of "interactivity" on a one-to-one basis came about early in the widespread adoption of the Internet and is still prominent today. However, as important as this concept may be, it has been studied in the past 15 or so years primarily from the social-psychological perspective with an implicit assumption of individuals socially isolated from one another. Concepts such as "perceived interactivity" have been developed and studied from the aspect of the single individual interacting with a medium and/or a message. Often, the study of interactivity has revolved around understanding the various psychological effects of Web sites' interactivity on an individual, for example.

While the study of individual differences with respect to interactivity has been helpful, inherent in this concept is the idea of, at minimum, two-way communication between sender and receiver. As noted by Rheingold (1993) and others early on in the development of the Internet, in addition to facilitating two-way communication, the Internet creates an environment where many individuals who are socially and geographically dispersed can *voluntarily* interact and collaborate with one another, which is exemplified by numerous online forums and virtual communities. That is to say, to the current time in academic study of this new medium, there has been little study of the ideas from the sociological perspective regarding the operation of human networks. Since the Internet is fundamentally a network, it seems clear that studying it must also involve the examination of social interaction with respect to this medium if a clear picture is to evolve about the full impact on individuals as well as on society at large.

This book is such a study, which is pioneering in this respect. Network is a concept with a relatively long history in the sociological discipline. This study is particularly notable in its application of some of these theoretical concepts to real-world problems as well as in the conclusions which have been drawn by the author about the nature of this new medium. Through an experiment involving the manipulation of two fundamentally different views of network structure, the author shows the impact of transforming network structures on important outcomes of communication effectiveness between senders and receivers of messages in this new medium.

This book should be valuable to read since it is one of the first academic attempts to shed light on how network structures influence the behaviors of individuals on the Internet. It is hoped that this study will spur further interest in the sociological investigation of this new and powerful medium.

John D. Leckenby
Professor and Everett D. Collier
Centennial Chair in Communication
The University of Texas at Austin

Preface

In the numerous articles and books about the impact of emerging new communication technologies, one thing is certain: the conventional divisions of communication are under severe attack. The boundaries between mass-mediated and interpersonal communication, advertising and public relations, and mass marketing and direct marketing are getting more and more blurred, and existing terms are losing their concrete meanings. If a person posts a message to recommend a product to many others, is this interpersonal or mass communication? If a company organizes a virtual community for promoting its brand/product, is this advertising or something else? Obviously, it is becoming more difficult to describe and cope with what is happening, let alone to explain why all of these things happen.

A better way to comprehend this foggy situation is, for a moment, to postpone struggling with the confusing terms, and instead to return to the basics—*people* who use the communication technologies—by asking a fundamental question: Why do we use the new communication technologies? Most previous studies had attempted to answer this question within the context of human-computer interaction (HCI) by focusing either on the characteristics of new media or on the perceptual processes of users. Hence, the reasons for using the new media were frequently reduced to the attributes of the media, such as convenience, interactivity, and flexibility. By fixating on the HCI, researchers overlooked the importance of *social need*—the need for connecting with and interacting with others—as a core reason people use new media. As pointed out by many scholars, when research is centered on technology, the people who use this technology are often overlooked.

Since then, the attention has gradually been shifted from dyadic HCI (I am not saying that HCI is not an important research topic) to multiagent interactions, thanks to the proliferation of *blogs* and social networking sites. Many studies dealing with virtual communities, blogging, online forums, and/or shared information databases have been published, and academic journals in communication and business have already published or are preparing to publish articles focusing on online social interactions and relationship building. The growing interest is not limited to academics, but extends to practitioners, and companies are beginning to recognize the enormous potential of online social collectivities for their businesses. By creating an online consumer forum for a product, for example, the company is able to interact with, disseminate information to, and get feedback from target consumers for a fraction of the traditional management cost. It is ironic that the nature of the Internet as the vastest social network ever is being rediscovered, despite the fact that it was born and has never stopped being a network.

The possibility of computer-mediated social interactions opens the door to serious challenges as well as to valuable opportunities for communication and business. The cornerstone of the many-to-many interactions is people's *voluntary participation*. Thus, the crucial question to ask is not just how to interact with individual consumers, but how to create and grow the online social collectivities in which a multitude of individuals voluntarily interact with one another. If everyone independently decides to participate, the aggregate outcomes will be simple sums of individual choices. Reality, however, is far more complicated than this. In an online social environment like virtual communities, individuals make decisions in conjunction with others' decisions, creating very complex social dynamics. How many people are in the community? How many of them are active players? How many of them are lurkers who benefit from the community without making their own contributions? Each person's motivation to communicate may be dependent on what others have done or are doing, and there may be numerous unknown factors affecting (positively or negatively) an individual's motivation. In this environment in which everyone responds to everyone else, a small and subtle change in local interactions can eventually determine the fate of the entire community.

Rediscovering the potential of the Internet as a computer-mediated social network leads us to very complicated, but intriguing, areas of inquiry into the social dynamics of multiagent interactions. The objective of this book is to illuminate the implications of the social aspect of new digital media on advertising, business, and communication in general. Understanding a social process requires examining both *micro* (intra-individual processes) and *macro* (aggregate of individual behaviors) levels, and how the two are interrelated. Scholarly attempts at linking these two levels of inquiry have been relatively scant in the fields of business and communication. I hope that this book will be a useful resource for those who want to bridge the two islands.

Acknowledgments

This book originates from my doctoral dissertation, and a derivative work of the dissertation has been published in the Journal of Communication. Work on this book would not have been possible without the help and support of the graduate faculty at The University of Texas at Austin and invaluable comments from many colleagues.

I am deeply thankful to my parents for their endless love and devotion, and also to my lovely wife, Hee-Jung, who has always been beside me and made my life rich and memorable.

Executive Summary

For the first decade after the introduction of the graphical browsers, the World Wide Web had been spotlighted as an ideal channel through which companies could engage in two-way communications with target customers. A number of articles and books, written either by academics or practitioners, supported such a view by characterizing it as an "interactive" medium, which would eventually *end* the era of mass marketing. These academics and practitioners, however, soon realized that the Internet would not wipe out all existing media, but instead would adapt to and find its place in the current media environment. New technology hype has gradually been replaced by a keen sense of reality, and managers in companies nowadays are far more conservative in evaluating the returns on their investments in interactive communication technology.

The Internet is now receiving another wave of interest because people have "discovered" that it is the vastest communication network ever built. As a matter of fact, the Internet was born a network and still is, but until recently its full potential had not been fully realized. In addition to facilitating two-way communication, the Internet creates a unique environment where multiple individuals can *voluntarily* interact with one another, as can be seen in online forums, shared databases, and virtual communities (Rheingold, 1993). Most groups or communities formed in this environment are *self-organizing* entities, which means that no central authorities for organizing or maintaining them are necessary. That is to say, virtual groups/communities are born and sustained through the voluntary communication efforts of many individuals.

This fascinating possibility of the many-to-many communication, however, leads us to unprecedented challenges of managing such online social collectivities. Human beings are often called *rational* actors who try to increase their gains/benefits, while minimizing losses/costs. (Perhaps this is why we all love free stuff!) Therefore, it is rational from an individual's standpoint to enjoy a product/service while paying minimal or no costs. Thus, a question that follows is: What would happen if everyone behaved rationally? The most famous and influential answer to this question was provided by Adam Smith, the father of modern economics: Pursuit of personal interests serves the common good. This idea is the cornerstone of the market economy in which the behaviors of egoists pursuing self-interests are coordinated by the "invisible hand."

An unfortunate fact is that the invisible hand sometimes fails. As Hardin (1982) puts it, "all too often we are less helped by the benevolent invisible hand than we are injured by the malevolent back of that hand" (p. 6). A public radio station, for example, provides its services for free to every listener, and it is rational for an individual to enjoy the service without donating money. If everyone does this, however, the public radio station may not be sustained for long, and no one will benefit from it any more, which is obviously irrational for everyone. This situation, in which individually rational decisions lead to a collectively irrational outcome, is called *social dilemmas* (Kollock, 1998; Komorita & Parks, 1996).

Similar problems can occur in online forums or virtual communities. The value of a virtual community depends on how much useful information is actively shared by the members. The more information is provided by the participants for sharing, the better for everyone in the group. From an individual's point of view, however, it is more rational to benefit from the information provided by other participants without making one's own contributions, because no one, including noncontributors, is alienated from the benefit of the group/community. If the majority of participants behave in this way, the information traffic will be substantially reduced, which would lower the value of the community. This special kind of social dilemma is called *communication dilemma* (Bonacich, 1990).

Many studies so far have been published to address the communication dilemmas in various contexts, and the suggested solutions fall into two categories. The first is to introduce a sanctioning system that gives individuals selective rewards and/or punishments based on their contributions. This solution, despite its merit, is limited in that there must always be an administrator who evaluates individuals' contributions and can offer rewards (mostly financial) and punishments. It is apparent that this solution is not viable for a purely self-organizing community with no central authorities or administrators. The second approach, on the other hand, focuses on individuals' motivation rather than on the reward-punishment relations. Scholars in this perspective have suggested that building a strong identity of a group or community may make individuals care more about collective than personal interests. Creating a solid collective identity for a virtual community, however, would be very difficult, if not impossible, especially when numerous anonymous individuals participate in the community.

An implicit assumption underlying previous studies of communication dilemmas is that there is only one possible form of group communication—*pooling* (e.g., an online public information pool shared by many people), in which individuals give their resources (e.g., information) to the public pool for sharing with everyone else, and people take these resources as they need. This form of group communication is particularly vulnerable to the communication dilemma, as discussed previously. Computer-mediated group communication, however, is not restricted to this particular structure, but may take other forms by connecting individuals in various ways, such as with dyad, triad, unilateral information relay, one-to-many communication, or a combination of them. A logical question that follows is whether individuals' communicative motivation in a virtual community situation varies depending on the forms/structures of group communication.

In order to answer this question, a longitudinal experimental study for comparing the performance of two communication structures—one involving a publicly shared online bulletin board and the other with multiple interconnected private boards (e.g., *blogs*)—was conducted. The former structure was named as the *group-generalized* exchange structure (G_{EX}) and the latter as the network-generalized exchange structure (N_{EX}). In the G_{EX} condition, experimental subjects exchanged information through a public online bulletin board, while each subject in N_{EX} was given a private bulletin board that had multiple hyperlinks to the boards of others in the group. The two experimental conditions were controlled in the same manner, except for their structures. Further, the performances of the two structures were monitored and compared in both small- and large-group situations to see if the structural change might affect individuals differently, depending on the group size. This experiment lasted for five consecutive days.

The findings reveal that N_{EX} produced a significantly higher level of individual contributions (e.g., posting information) than G_{EX} did. As expected, G_{EX} was found to be vulnerable to the free-rider problem, in which individuals' contribution level decreased slightly as the group size increased. This result seems consistent with the previous findings of the deteriorating effects of group size on cooperation, because the bigger the group gets, the less responsible individuals become for maintaining the group. As the group size increases, furthermore, individuals have fewer chances to make a noticeable difference, which lowers the efficacy of contribution. This is particularly true for G_{EX}, because all the messages posted by individuals are displayed together in a single bulletin board, which makes it difficult for one's contribution to stand out. In contrast, the level of contribution in N_{EX} increased as the group got larger. This is probably because individuals' responsibility for managing their own information boards should remain the same regardless of the group size, and their messages were shown in their own space separately from others' contributions, which made them more visible. In other words, N_{EX} not only preserved the degree of individual responsibility, but also enhanced the efficacy of making contributions, while G_{EX} lowered both.

The experimental findings imply two things. First, individuals' communicative motivation is dependent on the incentive structure of a particular form of group communication, which means that it is possible to influence individuals' motivation by altering the forms of communication. That is, people's willingness to share information can vary depending on the way the group communication is organized. Second, a network of private media, such as N_{EX}, is superior to a publicly shared medium, especially when a large number of individuals are involved. This implies that N_{EX} (networking individuals' private media) is more useful for organizing virtual groups/communities involving numerous anonymous individuals than is G_{EX} (grouping individuals through a public medium)—which explains partially the proliferation of *blogs*.

It is rapidly becoming clear that "interactive" communication is not limited just to the two-way communications between companies and consumers, but includes the dynamic interactions among consumers. An obvious (but often forgotten) fact is that people use the Internet more for communicating with other people behind the computer interface than with the computer or a company itself. What really matters at this point is not only how to interact with atomized individual consumers, but how to become an integral part of and coexist with the emerging *consumer network*. Hence, what advertising or marketing professionals need to understand is not only the dyadic consumer-computer/Web site interaction process, but also the nature of the consumer-generated network: How is it born? How does it grow and evolve? How, why, and under what circumstances do the elements of the network—consumers—flock together, interact, and cooperate? What factors facilitate or hinder such a many-to-many communication process? There are many intriguing questions waiting to be answered. What this book can show is only a small piece of a huge puzzle, but answering the questions lying ahead will lead us to a new level of understanding the ecology of online social collectivities, which would open the door to the untapped business opportunities.

References

Bonacich, P. (1990, June). Communication dilemmas in social networks: An experimental study. *American Sociological Review, 55*, 448–459.

Hardin, R. (1982). *Collective action*. Baltimore: The Johns Hopkins University Press.

Kollock, P. (1998). Social dilemmas: The anatomy of cooperation. *Annual Review of Sociology, 24*, 183–214.

Komorita, S. S., & Parks, C. D. (1996). *Social dilemmas*. Boulder, CO: West View Press.

Rheingold, H. (1993). *The virtual community: Homesteading on the electronic frontier*. New York: Addison-Wesley.

SOCIAL NETWORK STRUCTURES AND THE INTERNET

Introduction

> For all information's independence and extent, it is people, in their communities, organizations, and institutions, who ultimately decide what it all means, and why it matters.
>
> —Brown and Duguid (2000, p. 18)

Digital information technology, as is widely documented, has truly revolutionized the environment in which consumers search for, acquire, and utilize information. The Internet, the largest network ever created by humans, enables individuals to easily reach the largest storage of information through personal computers or handy wireless communication devices, such as PDAs or cellular phones. Visiting a retail store physically is no longer the only way to see the actual look of a product or to find the detailed information about it. In this electronically networked environment, instead, consumers can easily search or browse information provided by companies, observe available products displayed in three-dimensional forms, and even customize them based on their own preferences.

Advertising academics have agreed that the enhanced possibility of two-way interaction, which is captured by the concept "interactivity" (Leckenby & Li, 2000), makes the computer-mediated environment (CME) different from the traditional mass media environment. Communication in the mass media environment is largely unidirectional in the sense that a few dominating media corporations deliver standardized information and contents to *passive* mass audiences. The CME, however, creates an environment in which individuals are provided with a far more diverse set of choices and are enabled to interact virtually with an object or person (Schlosser, 2003) and to express their ideas and preferences directly to the companies. Stimulated by this technological change, most previous studies related to interactive advertising and electronic commerce have concentrated on understanding how the computer-mediated communication (CMC) environment influences the way consumers interact with companies (Kuk & Yeung, 2002; Rodgers & Thorson, 2000; Rust & Lemon, 2001).

Despite the impressive number of theoretical and empirical studies conducted to date, a cursory review of the existing literature of interactive advertising reveals that the single most important element has virtually been missed—*people*. In the previous studies, consumers are depicted as individuals who search for, acquire, and consume information, but not as *proactive* players who create and transmit information to others. Focuses have been placed on how consumers *react* to the stimuli associated with interactive features (Stewart & Pavlou, 2002), not on how they actively participate in communication with others in the CME. Forgotten is the role of people behind the dramatic transformation of the communication environment.

Overlooked is the obvious fact that tons of information flowing on the Internet has become available by the efforts of people like us. Ever since the potential of the Internet as an advertising medium was recognized, it has been veiled that consumers often use computers to interact ultimately with other consumers—the *social dimension of interactivity*.

Not surprisingly, it has often been the case that when people study technologies, humans behind them are shadowed and forgotten—technologies are understood as self-standing entities with their own rules and wills, which is never true. As Poole and DeSanctis (1990) aptly criticized: "Objectification and decontextualization conceal the social nature of technologies. Continually bombarded by such discourse, we forget that users constitute and give meaning to technologies" (p. 178). The actual value of a communication technology is not derived from its inherent functional characteristics, but from the ways people use it and attach meanings to it.

As Morris and Ogan (1996) appropriately put it, the Internet is "a multifaceted mass medium...[that] contains many different configurations of communication. Its varied forms show the connection between interpersonal and mass communication that has been an object of study since the two-step flow associated the two" (p. 42). In this networked environment, the boundary between mass-mediated and interpersonal dimensions of communication is blurred. In the traditional local social communities, people with whom individuals can communicate interpersonally are limited to those geographically proximate or with prior social relationships. However, the CME allows individuals to communicate even with complete strangers who are socially and geographically distant (Wellman, 2001).

The value of the Internet cannot be derived from its technological capacity itself, but from how people use the medium for their communication activities. As Wellman and Hampton (1999) put it: "When computer networks connect people and organizations, they are the infrastructure of social networks" (p. 649). The Internet can be understood as a computer-mediated *social network*, which visualizes the revolutionary shift "from living in 'little boxes, to living in networked societies" (p. 648). What should be noted here is a shift from *closed groups* or *local social clusters* to global networks opened to the world outside, whereby people communicate with one another beyond prior social boundaries (Wellman et al., 1996).

As Granovetter (1973) noted earlier, new ideas and information are often communicated through *weak* social connections (e.g., mere acquaintances) rather than *strong* ones: *Homophilious* or close-knit groups "act as a barrier preventing new ideas from entering the network," while *heterophilious* or sparsely networked groups enable "innovations to flow from clique to clique via liaisons and bridges" (Rogers, 1976, p. 299). This theoretical notion clearly epitomizes the impact of the CME. Facilitating communication among strangers may increase dramatically the probability that a piece of information is transferred from one social cluster to another, which remarkably accelerates the speed of information diffusion. By allowing individuals to communicate with multiple others beyond social and physical boundaries, in short, the CME creates a unique environment where the impact of the information disseminated by an individual consumer becomes comparable to that of a large corporation. It is of no doubt that this expansion of the range of consumers' information transmission may have a direct impact on the performance of products, services, and even companies in this new communication environment. This is why advertising scholars and practitioners should pay close attention to the convergence between interpersonal and mass-mediated communication.

Virtual Communities and Communication Dilemma

Virtual communities or *online feedback systems*, which are already ubiquitous on the Internet, are the eminent examples of the environment facilitating *many-to-many* information exchange online (Hagel & Armstrong, 1997; Muniz & O'Guinn, 2001; Rheingold, 1993). A common characteristic of such virtual social gatherings is that each individual's voluntary communication activity is essential for creating and maintaining the collective entities (Bagozzi & Dholakia, 2002). Unlike the traditional local communities that are geographically defined (e.g., neighborhood) or formed based on preexisting social relationships (e.g., friendship, kinship), virtual communities are sustained only by individuals' *volitional* communication behaviors. Typically, a virtual community is operated based on a shared communication channel, such as an electronic bulletin board or a group mailing list. Thus, an individual's presence and participation in an online community can become visible only by the messages posted by the person. Since no one can force others to be more cooperative, this communication environment is often characterized as *decentralized* (Rheingold, 1993).

In this decentralized environment where autonomous individuals interact with one another, a dilemmatic situation may occur. In a typical virtual community setting, which is based on an electronic bulletin board, individuals may think it unnecessary to make their own contributions because there are already many others who have made and will continue to make contributions to the community. Even if they do not make their own contributions, individuals can benefit from the contributions made by others. Since there is no discernable difference in benefit between people who contribute and those who do not, people may be strongly tempted to *free ride* on others' contributions. If everyone wants to free ride (rational individual decision) in this situation, however, it is evident that the virtual community cannot be sustained, and eventually no one will benefit from the community (collectively irrational outcome).

Social scientists have called this situation, in which individually rational decisions lead to a collectively irrational outcome, a *social dilemma* (Kollock, 1998; Marwell & Ames, 1979). Suppose a classic music program is aired by a public radio station, which is maintained by the donations of listeners. Because they are not required to give donations to enjoy the program, individuals may not have any incentive to donate to it. If every listener decides not to make a donation, the public radio station cannot be sustained, and will not be able to provide the music program anymore, which is surely not the consequence that everyone expects. Actually, we may easily find similar examples, such as environmental conservation or constructing a public library in a neighborhood. Because a commonality underlying these situations is that various "public goods" are involved, they are often called *public good dilemmas* (Messick & Brewer, 1983). In reality, a considerable proportion of virtual communities on the Internet are suffering from the lack of people's contributions, which results from similar dilemmatic situations. Bonacich (1990) called this particular type of public good dilemma, which occurs in a group communication situation, a *communication dilemma*.

By treating a virtual community as a group entity in which participants are fully committed to the group's goal (see Bagozzi & Dholakia, 2002; Granitz & Ward, 1996; Hemetsberger, 2002; Muniz & O'Guinn, 2001), past research has neglected the existing tension between individual and collective interest in virtual communities. Scholars studying virtual communities have been overly optimistic about the success of creating and maintaining the computer-mediated social groups, even when most virtual communities suffer from the *suboptimal* situation in which most members in the group free ride except for a few dedicated contributors. Because of this neglect of the individual-collectivity tension, communication researchers have portrayed consumer word of mouth (WOM) and diffusion as a *conflict-free* process in which microinteractions among individuals translate into macroscale outcomes without a problem (Bonacich, 1990).

How can the problematic situation be resolved? Social dilemma researchers have found that the dilemmatic situation commonly results from the two characteristics of public goods—*nonexcludability* and *jointness of supply* (Komorita & Parks, 1996). Nonexcludability, first, means that no one can be excluded from the benefit of public goods once these goods are provided. In a virtual community, the information provided by a person can be accessed and used by anyone within the community regardless of not making contributions. Second, jointness of supply means that an individual's consumption of a public good does not reduce the proportion for others' consumption. Because the amount of utility that consumers can obtain remains the same, this implies that an individual's exploitative behavior does not cause direct conflicts with others. Due to this incentive structure, an individual may be easily tempted to behave exploitatively, lurking behind the contributions of others.

As a solution to the free rider problem in virtual communities, this research proposes to transform a public communication space into *a network of multiple private spaces*. Social exchange theorists have suggested that actors' motivation to cooperate or defect is contingent on the *structure* of resource exchange given

(Emerson, 1976). Based on the theoretical tradition, for example, Yamagishi and Cook (1993) and Molm (1994) asserted that there should be some structures of resource exchange, which motivate individuals to cooperate with others, while others leave them to free ride. In this line of thought, this study attempts to combine two key strategies—(1) *privatization* and (2) *networking*—to remove the structural deficiency underlying typical virtual communities. The primary objective of this research is to examine empirically how the varying structures of online information exchange influence the levels of consumers' cooperation in a virtual group setting.

Chapter 1 provides a general background for illuminating the nature of the Internet as a networked environment. In this chapter, the current status of interactive advertising research is briefly reviewed, and the social dimensions of interactive media, which remain relatively unexplored, are illuminated. Particular emphasis is placed on discussing how the concept of "network" can be applied to understand the full potential of the Internet as an advertising medium, and to understand the impact of the networked communication environment on the patterns of information diffusion.

Chapter 2 attempts to bridge the two traditions of research—the studies of *diffusion process* and *social exchange* theories—to provide a theoretical perspective to illuminate the various issues arising from *many-to-many* information exchange situations in the CME. The chapter subsequently discusses in detail how an individual's motivation can be in conflict or coordination with others in multiagent communication settings. Based on the literatures of game theory, social exchange, and collective action processes, the potential problems of organizing virtual groups or communities are delineated, and a structural solution for the problems is suggested.

In chapter 3, the hypotheses to be tested are drawn, based upon the discussions of the previous chapter. In chapter 4, the procedures and results of two pretests (which were conducted prior to the main experiment) are presented. Building upon the pretest results, the experimental and measurement procedures of the main test are then illustrated in detail. Chapter 5 presents the findings of the main experiment and data analyses procedures, and chapter 6 discusses the results found and the relevant issues in more detail. Finally, chapter 7 discusses the implications of this study's findings for future research as well as for managerial issues of interactive advertising and online relationship building with potential consumers.

CHAPTER 1

SOCIAL DIMENSION OF INTERACTIVE MEDIA[1]

> What makes a social system, in contrast to a set of individuals independently exercising their control over activities to satisfy their interests, is a simple structural fact: Actors are not fully in control of the activities that can satisfy their interests, but find some of those activities partially or wholly under the control of other actors.
>
> —Coleman (1990, p. 29)

A widespread tenet in advertising research, which is ultimately a misunderstanding, is that each consumer *independently* evaluates and decides to purchase a product based on his or her personal needs and interests. Reality is far from this belief. In fact, a new product is introduced to an individual consumer *embedded* into ongoing social relationships (Granovetter, 1985), in which the person's needs and interests are inseparably related to those of others. An individual's adoption of new ideas or products is often a response to the needs arising from his or her social environment, not purely to one's own interests (Leibenstein, 1976). In the real world, it is a ubiquitous phenomenon that the probability of a person's adoption of a new product changes, depending on how many others within his or her immediate social environment already adopted it (Granovetter, 1978). It is in this sense that a product is introduced not to individuals in isolation, but to *networks of individuals*.

Scholars have called this—the reliance of an individual's adoption of a product on the proportion of other consumers who already adopted it—"network externality" (Katz & Shapiro, 1985). *Network externality* refers to a situation in which the value a person obtains from consumption of a good "increases with the number of other agents consuming the good" (p. 424). Why does the network externality exist? Network externalities or network effects occur when the value of a product is dependent not only on its own functional utility, but also on the whole properties of the social network of consumers using the product. Noting the networks of consumers rather than consumers *per se* is particularly important when evaluating the value of a new communication technology, because the value of a new communication technology is derived not from its functional benefit for each individual, but from its impact on the way people communicate with one another.

The value added by a new communication technology is judged in terms of the characteristics of the preexisting social communication environment. For example, an individual's adoption of the telephone cannot be explained fully by his or her personal needs, because the needs arise from the individual's social surroundings—how many of his or her friends or adjacent households currently own telephones. The more people that own telephones, the more beneficial it would be for a person to obtain one. Numerous similar examples like a fax machine, e-mail account, and cellular phone can easily be found in the real world.

For the past decade, the Internet has been spotlighted as a new medium enabling *two-way* communications between companies and target consumers. A considerable number of studies have been completed and published, in order to grasp the unique effects of this "interactive" medium on the traditional marketing communication activities. The term, interactivity, has rapidly become part of an essential vocabulary in Internet research. Fixing our eyes on the human-medium interaction, however—the most important characteristic of the Internet—has slipped away from our sight: the Internet is, by definition, a *network*. The Internet is the largest and fastest-growing network of computers in the world, through which people, organizations, and even countries can interact and communicate with one another. In this sense, the Internet can be considered a *computer-mediated social network* (Wellman et al., 1996).

The objective of this chapter, as a general background of this research, is to point out that the key feature of the Internet, which differentiates its from other traditional media, is its revolutionary way of connecting individuals and organizations with one another—not only its technical features enabling one-to-one (dyadic) interactions between a person and a medium. Recognizing the potential of the social dimension of the computer-mediated network is essential for successful marketing communication in this new environment. In the following sections, previous theoretical and empirical studies regarding interactive advertising conducted up to this point are briefly described, and the implications of the Internet as an expanding network for market information diffusion are discussed.

The Current Status of Interactive Advertising Research

It has already been widely documented that the Internet as a new medium has fundamentally transformed the traditional source-oriented advertising paradigm, and that the concept of *interactivity* is at the core of this change (e.g., Leckenby & Li, 2000; Pavlou & Stewart, 2000; Rodgers & Thorson, 2000; Stewart & Pavlou, 2002). The uniqueness of the Internet for advertisers emanates from its ability to locate ordinary people at the center of the communication process (Pavlou & Stewart, 2000; Stewart & Pavlou, 2002). Unlike the traditional mass media, which are owned and controlled by a few media corporations, the Internet has been constructed and sustained by the collective efforts of ordinary individuals.

Recognizing the potential of this interactive medium, advertising academics have studied how this new communication medium alters the way in which companies interact with target consumers. Cho (1999) grouped interaction occurring in the online environment into three different categories: (1) *human-human interaction*, (2) *human-machine interaction*, and (3) *human-message interaction*. Communication scholars with interests in CMC research have dealt with the first type of interaction (e.g., Walther, 1996). In contrast, prior interactive advertising research has focused primarily on the second and third types of interaction. The main research question has addressed how an interactive medium like the Internet facilitates the interactions between companies and consumers.

The studies conducted thus far can be roughly grouped into two distinct lines of inquiry. The first, often called the *mechanical approach* (Ha & James, 1998), views interactivity as a functional attribute of a medium or message vehicle (e.g., Web site), and examines the effects of various interactive features on consumer's information processing and decision making. For example, Steuer (1992) defined interactivity as "the extent to which users can participate in modifying the form and content of a mediated environment in real time" (p. 84); the levels of interactivity are determined by the characteristics of a medium used. As such, this mechanical perspective places the focus of research on how the different configurations of a medium's interactive features affect consumers' responses to commercial messages.

Following this approach, many interactive advertising studies have been conducted to investigate the effects of advertising messages associated with the interactive technologies of the Web (e.g., banner ads, product/corporate sites) on consumers' various psychological responses. Some studies attempted to answer how the varying levels of interactivity of online banner advertisements affect consumers' attitudes toward the brand and the advertisement (i.e., Cho & Leckenby, 1999; Cho, Lee, & Tharp, 2001; Li & Bukovac, 1999), as well a how they affect click-through rates (Lohtia, Donthu, & Hershberger, 2003).

Beyond the banner advertisements, Ghose and Dou (1998) examined the effects of interactive components of Internet presence sites (IPS) on their appeals, and Coyle and Thorson (2001) investigated, in an experimental setting, the effects of the varying levels of interactivity and *vividness* of Web marketing sites on consumers' perceived *telepresence*,[2] on their attitudes toward the site, and on the levels of attitude-behavior consistency. Fortin and Dholakia (2005) also examined the effects of interactivity and vividness of Web advertisements on consumers' attitudes and behavioral intentions. In a similar vein, by manipulating the amount of user control allowed, Ariely (2000) investigated how varying levels of user control of the information-search system influence consumers' memory, confidence, and the quality of decision made on the information found. Further, Macias (2003) attempted to build a structural equation model to understand the effects of the interactivity of a Web site on consumers' levels of message comprehension and attitudes.

In contrast to the mechanical approach, some scholars have emphasized the role of users' perceptual processes rather than the functionality of a medium (Newhagen, Cordes, & Levy, 1995). By noting that interactivity is inherently a subjective experience in the eyes of the beholder, researchers in this perspective have attempted not only to investigate the internal structure of the psychological experience—*perceived interactivity* (McMillan & Hwang, 2002; Wu, 2000)—but also to identify possible factors affecting both it and its psychological effects (McMillan, Hwang, & Lee, 2003). In sum, the studies in this line of thought have weighted individuals' personal characteristics and motivation more than the functionality of a medium.

It is still under academic debate whether the concept of interactivity should be regarded as an attribute of a medium, or as a person's subjective perception. For example, Sundar (2004) argued that interactivity should be considered a technological variable, because an individual's interactivity perception is an outcome of interactivity, not an independent cause. Building upon Rafaeli's (1988) classic unidimensional definition,[3] he suggested that interactivity should be defined as the degree of *sequential responsiveness* of a medium or of message like a Web site with a hierarchical link structure. In contrast, Bucy (2004) proposed that interactivity is best defined as a perceptual variable because it reflects individual variations in experiencing technology-mediated communication.

Medium-in-Interaction: A Sociological Turn

After all the research efforts over the past decade, it is ironic that we are left with a pile of questions instead of reliable answers. Is interactivity a trait of a medium? Does it refer to just a psychological outcome of HCI, or should the concept reflect the dynamic process of interaction? Before trying to define the construct, let us return to the most fundamental question that has not been answered satisfactorily: Why should we study interactivity? The concept of interactivity has been spotlighted, thanks to the common notion that it is the key factor in differentiating the Internet from traditional mass media: unlike other mass media, the Internet allows companies to communicate bidirectionally with their potential customers. This statement is true particularly from an advertiser's standpoint, because the enhanced possibility of making two-way communication with consumers is the major reason why corporations have rushed into cyberspace.

Then, is this possibility of two-way communication with companies also the major reason for consumers to use the medium? Do consumers always get online and surf the Web in order to interact with companies? What if consumers have a different motivation for surfing the Web than advertisers expect they do? Indeed, social interaction entails, by definition, *mutual involvement* between actors (Goffman, 1957). Understanding how the goals and motivations of two interacting entities are interrelated and coordinated, therefore, should be the basic ingredient for making the interaction possible (Stewart & Pavlou, 2002).

Viewed only from the managerial perspective of advertisers, the consumer-medium (or message) interaction has been reduced to the fragmented action/reaction between an individual consumer with information needs and a medium like an information vending machine responding to the user's input. As a result, previous interactive advertising studies have lost sight of the other side of a consumer's motivation for using the Internet: people get online not only to search and acquire information provided by commercial/noncommercial organizations, but also to interact with other people (Wellman, 2001).

An interaction between a human and a medium is not an event occurring in a *social vacuum*. As the concept of network externality implies, an individual's motivation to adopt/use a communication medium is inseparably related to the contextual characteristics of his or her social environment (Fulk, 1993). In other words, understanding the motivation underlying one's usage of communication media requires considering the social, economic, and cultural context in which the person is embedded. When using a communication medium, the individual may actually aspire to interact with other people or organizations that share certain relationships or interests with him or her (Bargh & McKenna, 2004).

Stromer-Galley (2004) pointed out that it is important to distinguish two different types of interactivity: (1) interactivity as a *product*, and (2) interactivity as a *process*. The former is related to the context of human-computer (or message) interaction (HCI), while the latter is related to computer-mediated human-human interaction (CMC). She argued that the two interactivity types should be treated separately, because they are different phenomena, which cannot be labeled by a single name. This is a timely and reasonable claim, given the existing conceptual confusion of interactivity. However, is it all we can do merely to leave them distinguished, even though both of them are the crucial dimensions reflecting the potential of the Internet? As a matter of fact, what should be noted here is that the explosive potential of the Internet as a communication medium is based on the intersection of the two interactivity types.

There is growing evidence showing that people access the Internet not only to obtain a response from a *nonhuman entity* like a computer or a Web site, but also to exchange contents and information with the people or organizations behind it. Prominent examples are virtual communities and blogs (Web-logs), through which individuals are freely engaged in communicating with multiple others. In this situation, understanding consumers' motivation to be part of the computer-mediated social network requires seeing a bigger picture encompassing multiple actors hooked up to the network, as shown in Figure 1.1. Evidently, a dyad (HCI) is just part of a bigger network, and the myopic view focusing exclusively on a dyadic interaction between a human and a medium may lead to losing sight of the entire network.

This seemingly straightforward notion, however, has not been clearly recognized. The early entrepreneurs on the Internet focused on the capacities of this medium to deliver audiovisual

FIGURE 1.1. Dyad-in-network.

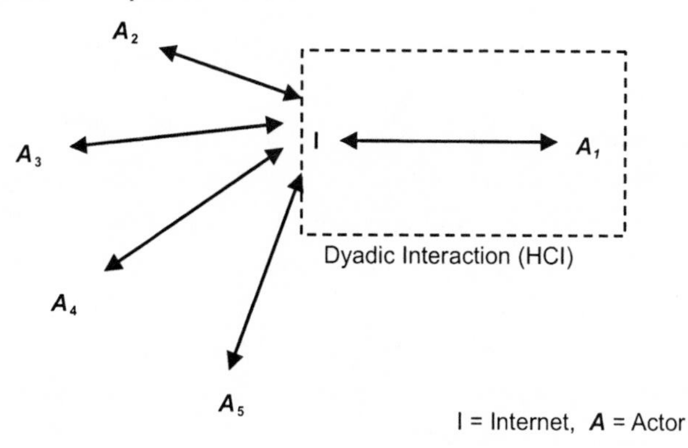

contents in real time to consumers, and made a substantial investment in developing "interactive television" and similar technologies (Bargh & McKenna, 2004). Even though the usefulness of the real-time delivery of customized media contents is evident, concentrating only on those features has made these entrepreneurs overlook the tremendous implications of the social dimension of the Internet on the current marketing communication environment. Without seeing the vast networks beyond a dyadic interaction between a person and a medium, the full potential of the Internet as an advertising medium will never be revealed.

Bridging Interpersonal and Mass-Mediated Dimensions

Market information in this CME can travel through consumers' *peer-to-peer* communication channels much faster than it can anywhere else. In the traditional mass-media environment, consumer WOM is largely bounded by social and physical constraints: individuals normally communicate product-related information with others within their social circles or clusters, but rarely with strangers or mere acquaintances, which minimized the impact of consumer WOM on the market process. The Internet has, however, changed this situation. Consumers can now have product-related conversations with numerous others who are not socially or physically proximate, and an individual's transmission of positive or negative information regarding some products or companies may have an enormous impact on their performance in the market. Given this situation, advertising academics and practitioners should think about how to use the expanded channels of consumer social interactions for marketing communication purposes.

As mentioned before, there can be two different types of interactivity—(1) human-medium/message interactivity, and (2) human-human interactivity. For convenience, let us call the first one Type I, and the second one Type II. By combining the two types of interactivity, we can draw a map consisting of four quadrants as seen in Figure 1.2.

Figure 1.2 illustrates a map, in which a variety of advertising formats currently in use are differentially positioned based on two criteria: Type I and Type II interactivity. As mentioned before, Type I interactivity indicates the degree of interactivity between the consumer and a medium (or message), while Type II interactivity refers to the extent to which consumers interact socially with one another. Traditional advertising forms like television and newspaper advertising are positioned in the third quadrant, since they involve neither two-way interactions between consumers and products/companies nor consumer-to-consumer social interactions. Consumer word of mouth (WOM), by contrast, is located in the second quadrant, indicating that it is high in Type II interactivity, while low in Type I. Consumer WOM has

FIGURE 1.2. Mapping interactive advertising types.

Low ◄------- Type I -------► High		Type II
• Consumer WOM • Multilevel Marketing	• Virtual Communities • Blog Advertising • Mobile Advertising	High
• Broadcasting Advertising • Print Advertising	• Corporate Web Site • Advergaming • Banner Advertising	Low

been recognized as a powerful marketing tool, though it is inherently a social phenomenon and thus is not under the control of companies (Dichter, 1966).

Compared with the old advertising formats, Internet advertising (including banner ads and corporate Web sites) is placed in the fourth quarter, which means that it is high in Type I and low in Type II interactivity. Existing interactive advertising forms have been developed and used primarily to enhance the level of interaction between consumers and advertising messages associated with interactive features. For example, *advergame*—the combination of advertising and online gaming—is an advertising form for motivating consumers to interact with various verbal and visual messages, which may increase brand awareness and attitude (Hernandez, Chapa, Minor, & Barranzuela, 2004). Although advergame may take other forms (like network games), involving multiple individuals simultaneously, most of these advertising forms are not intended to relate advertising with consumer social interactions.

The only quadrant remaining relatively unexplored thus far is the first one, which requires considering the joint increase in both Type I and Type II interactivity. It has become increasingly evident that broadcasting commercial messages to consumers is no longer the only viable way to disseminate product information, increase brand awareness, or strengthen the image of a brand in the online environment. The proliferation of virtual communities demonstrates that leaving consumers by themselves to talk about brands, products, and companies may be a remarkably cost-effective way for companies to get better-than-expected outcomes (Hagel & Armstrong, 1997). As is widely acknowledged, this digital social collectivity, which is not geographically defined, enables countless far-flung individuals to interact and communicate a variety of product-related information and opinions, which forces companies to escape from the conception of the "mass" market that consists of atomized consumers.

How can firms use the digital social collectivity as a potential channel for marketing communications? First, companies can be the direct *organizer* of the virtual consumer communities (Balasubramanian & Mahajan, 2001; Hagel & Armstrong, 1997). By organizing and maintaining a virtual community, companies can not only obtain customer loyalty by interacting closely with potential consumers, but they can also deliver to potential buyers, at extremely low costs, a variety of information regarding their products and services. Moreover, since the virtual consumer communities can function as a useful channel for marketers to hear potential consumers' opinions on their products or on companies in general, it may be used for testing a new product before introducing it into the market.

Secondly, companies may advertise their products and services to the virtual communities, since they provide the most accessible routes to the networks of consumer WOM. A recent attempt of this sort is the "blog-ad," which refers to placing advertisements on personal blogs. A blog is a personal homepage with more enhanced connectivity than an HTML-based Web site. In the blog system, individuals' communication activities (e.g., writing and posting messages) automatically form networks through which they can move from one blog to another without manually adding hyperlinks to each page. Once a communication network is formed, information may flow easily through it. That is, the blog-ad is an attempt to expose the information of products and services to consumers interconnected by the networks.

Development and rapid diffusion of mobile communication devices expands the range of interpersonal communication beyond virtual communities. Nowadays, consumers with mobile communication devices are able not only to exchange verbal and visual information with their friends efficiently, but also to disseminate the information obtained to many others by posting it on the Web. This means that the clear distinction between the offline and online environment becomes hazy, for they are rapidly being integrated. This changing media environment epitomizes the importance of the social dimension of interactivity and the direction to which interactive advertising should move.

MEDIA EVOLUTION AND CHANGING COMMUNICATION ENVIRONMENT

Where are we now on the trajectory of media evolution? We may start this discussion from Hoffman and Novak's (1996) conceptual representation of three different forms of communication: *one-to-one*, *one-to-many*, and *many-to-many* communication.

FIGURE 1.3. One-to-many communication model.

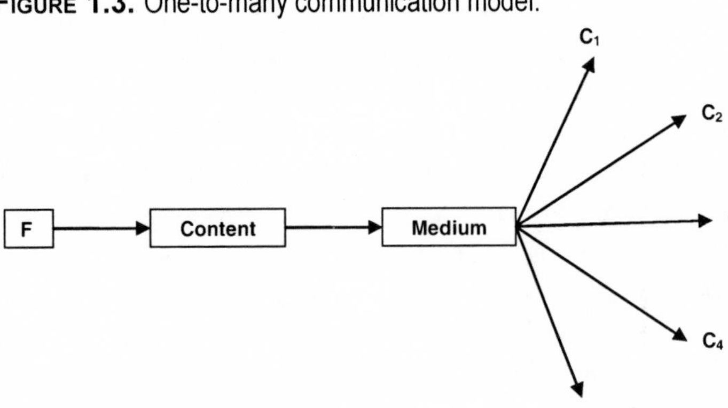

Source. Hoffman and Novak (1996).
Note. C = Consumer, F = Firm.

In their conceptualization, the one-to-one communication model represents a *dyadic* interaction situation, in which two individual entities communicate with each other either face-to-face (nonmediation) or through a medium (mediation). In this form of communication, a person's motivation and behavior are contingent on the interaction partner's attributes and behaviors, as shown in the evidence found in social psychology (Fiske & Taylor, 1991). Previous interactive advertising or other Internet-related studies have focused on this type of interaction, particularly between an individual and a message mediated by a technology.

The *one-to-many* communication model (Figure 1.3) represents a mass communication situation, in which an identified source disseminates messages to many recipients simultaneously. In this model, a source dominates the power to control the communication process, while recipients are portrayed as passive; the message flow from a source to a multitude of recipients is *unidirectional*, which means that reciprocation between the two is very limited (Rubin, 1994). This form of communication is exemplified clearly by the *broadcasting media*, such as television, radio, or newspaper, with which audiences can only watch or listen to visual and/or verbal messages delivered from a few sources. This model has been widely accepted as the dominant model for marketing communications because it is a cost-efficient way to deliver standardized messages simultaneously to countless mass audiences

The *many-to-many* communication model (Figure 1.4) represents the Internet, which is defined as "a dynamic distributed network, potentially global in scope, together with associated hardware and software to access the network, which enables consumers and firms to 1) provide and interactively access hypermedia content ('machine interactivity') and 2) communicate

FIGURE 1.4. Many-to-many communication model.

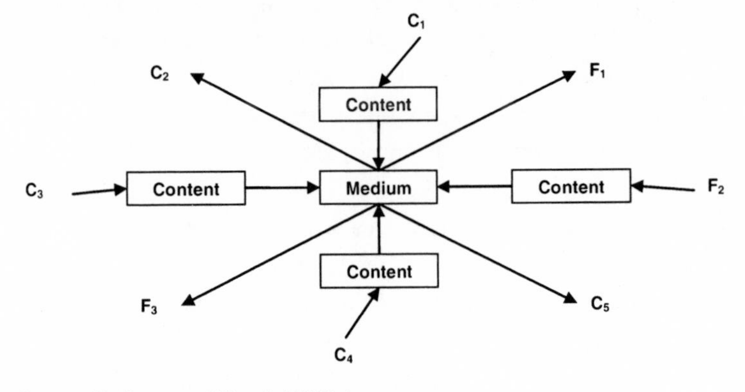

Source. Hoffman and Novak (1996a).
Note. C = Consumer, F = Firm.

through the medium ('person interactivity')" (Hoffman & Novak, 1996, p. 53). Scholars have noticed these two unique characteristics of this many-to-many communication situation.

First, the *role structures* of communicators (e.g., information sender or receiver) are not fixed as in the one-to-many communication model, but become flexible. This means that it is no longer predetermined whether a person in this environment can be an information sender or a receiver. As a matter of fact, a considerable portion of the contents on the Web are provided by ordinary people through personal Web sites or electronic bulletin boards.

Second, this model integrates one-to-one and one-to-many modes of communication. Here, an individual can not only communicate with another individual on a one-to-one basis, but also can broadcast messages to numerous others simultaneously. For this reason, Hoffman and Novak (1996) distinguished *person interactivity* from *machine interactivity*. We may easily find on the Web that there are numerous individuals trying to disseminate

FIGURE 1.5. Many-to-many communication in mobile communication environment.

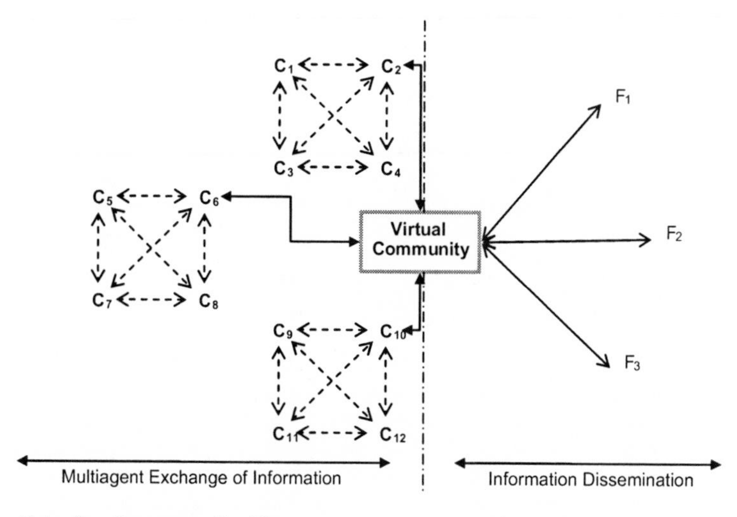

Note. C = Consumer; F = Firm.

and find useful information through various channels like bulletin board systems (BBSs). In this circumstance, mass advertisements cannot be dominating information sources; they must compete with numerous "individual broadcasters" to deliver appropriate product-related information to target populations.

The recent spread of mobile and wireless communication devices blurs the boundary between the online and offline environment and accelerates the integration process of the two. Imagine that a consumer has a cellular phone with a digital camera function. This individual can easily take a picture of a new laptop computer or a nice restaurant and share it with friends by sending it directly or uploading it to his or her blog homepage. The information uploaded to one's blog is accessible not only to the individual's direct contacts, but also to others who have *indirect* connections with the original information source. Through a nexus of individuals' information exchanges, the information can be diffused widely and quickly.

Figure 1.5 illustrates the situation described previously. The widespread use of mobile communication devices connects the two sectors of communication—online and offline—which had previously been thought separate from each other. Interpersonal communication within local social networks (e.g., friends) is facilitated by the use of mobile communication devices, and marker information circulated in a social cluster can be transferred to another social cluster through the virtual community environment. This means that the interpersonal communication networks among consumers are becoming more accessible to companies.

Network, Communication, and the Flow of Market Information

In this networked communication environment, the number of potential recipients who may be exposed to a message transmitted by an individual may increase *exponentially*, because many different local social networks that are geographically or socially far apart can be connected. As a result, the sector of decentralized communications among consumers becomes more dynamic and influential.

According to Granovetter (1973), diffusion is a process through which "small-scale interaction becomes translated into large-scale patterns, and that these, in turn, feed back into small groups" (p. 1360). In his account, what makes this micro-macro interaction possible is the *weak social tie*, which plays an important role in connecting or bridging separate social clusters: "whatever is to be diffused can reach a larger number of people, and traverse greater social distance, when passed through weak ties rather than strong" (Granovetter, 1973, p. 1366). This means that large-scale information diffusion becomes possible mainly through weak social ties serving as the bridges between social clusters.

Diffusion researchers have long accepted this sociological notion and found empirical evidence confirming this proposition, that a *heterophilious* community facilitates the process of new information diffusion better than does a *homophilious* group (Brown & Reingen, 1987; Rogers, 1976, 2003; Rogers & Bhowmik, 1970; Valente, 1995). When people need new information or advice that cannot be obtained directly from close friends or family, they often consult with and are helped by others who are not socially close to them. For example, Granovetter (1974) found that people often acquire job information from those who are casual acquaintances, not from close friends or family members. Also, Constant, Sproull, and Kiesler (1996) found that people in an organizational setting used a computer network as a channel to communicate with strangers to obtain necessary information and advices.

The most significant structural change of the media environment is that the *bridges* connecting people who do not have any prior social relationship has become ubiquitous—the probability that a message goes beyond a local social boundary increases dramatically. Before the advent of the CME, a verbal message delivered through word of mouth usually circulated among socially or geographically close individuals and rarely reached people outside of the circle. In contrast, a message posted on an electronic bulletin board can easily be seen by individuals who are not part of the social system of the original information provider.

Theoretically, this drastic change in communication structure among individuals may correspondingly cause a significant alteration in the *patterns of information diffusion*. Diffusion theorists have identified two distinct shapes of the diffusion process: *exponential*, with only decreasing marginal returns, and *sigmoid*, or S-shape. The sigmoid shape of the diffusion process, which has been found as the most typical pattern of diffusion by empirical research, is an S-shaped (or logistic) curve showing a slow increase in an initial phase followed by an explosive increase at a *critical mass* point, which in turn is followed by decreasing marginal returns in later stages (see Figure 1.4).

Why do the sigmoid patterns occur? Summarizing the previous diffusion-related literature, Gatignon and Robertson (1985) mentioned that the sigmoid pattern of diffusion is based primarily on "the concept of social imitation or personal influence" (p. 859),

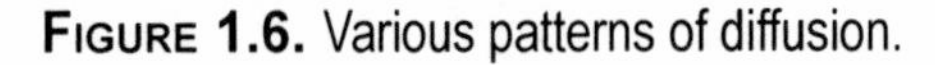
FIGURE 1.6. Various patterns of diffusion.

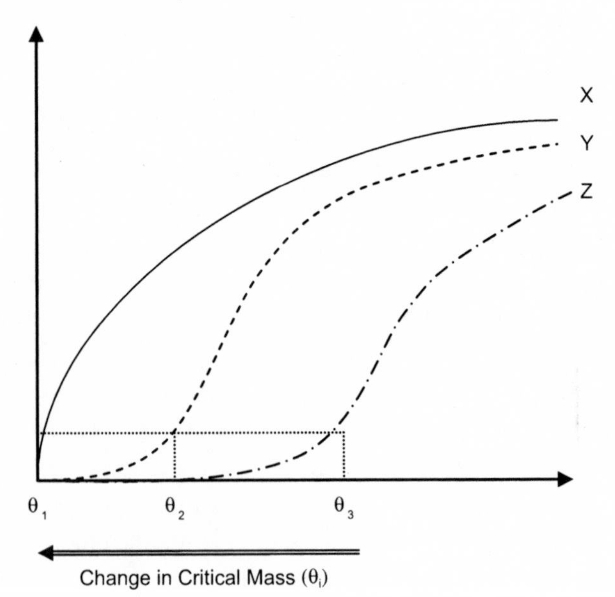

among many relevant factors. That is, the slow increase in the initial phases in the sigmoid pattern reflects the *structural constraints* imposed on interpersonal communication through word of mouth. Because an individual is able to send a message to only a few socially reachable others, it should inevitably take a long time for the message to reach enough people.

The second diffusion pattern identified is the *exponential* function. Holding other factors (e.g., the level of uncertainty, the number of learning requirements) constant,[4] diffusion theorists have postulated that this exponential pattern of diffusion tends to occur due to "a relative lack of personal influence" (Gatignon & Robertson, 1996, p. 859). The exponential curve, with decreasing marginal returns, illustrates a pattern: the rate of increase is very high in the first phase, but decreases continuously over time. As illustrated in Figure 1.4, the exponential function (graph X) has a shape showing a continuous decrease in the rate of marginal increase. The shape of this diffusion curve may reflect a situation in which information is delivered to most people primarily by mass media, with minimal interpersonal mediation.

Given the radical transformations in communication structures made possible by the Internet and mobile communication technology, it may be inferred that the exponential pattern of diffusion may occur not only because of a lack of interpersonal mediation, but also because of the enhanced dynamics of interpersonal interactions. That is, the multidirectional diffusion of information beyond one-to-one transmission (e.g., word of mouth) may make it easier to surpass the level of *critical mass*[5] of information diffusion, which means that a message or information transmitted by a person can become visible to everyone much faster than ever before. As illustrated in Figure 1.6, the rapidly expanding communication networks may ultimately move the time taken to reach a critical mass from θ_3 to θ_1, which eventually will result in a change in diffusion patterns from Z to X.

Undoubtedly, recognizing the impact of the convergence of interpersonal and mass-mediated dimensions of communication on the speed and range of information diffusion has an important implication for marketing communication on the Internet. This requires stepping away from the conception of the mass market as *interpersonal vacuum* (Ben-Porath, 1980) and developing a new perspective that is appropriate for grasping the dynamics of social interactions among individual consumers. In short, this change in communication structure, triggered by the advent of interactive communication technology, compels us to shift our attention from the fragmented attempts to interact with individual consumers to the systematic interventions into the *decentralized social communication process*, in which voluntary behaviors of individual consumers (e.g., information transfer) play a central role.

Endnotes

1. The previous version of this chapter was presented in a preconference session at the 2004 annual conference of the American Academy of Advertising in Baton Rouge, Louisiana.
2. *Telepresence* is defined as "a psychological state or subjective perception in which even though part or all of an individual's current experience is generated by and/or filtered through human-made technology, part or all of the individual's perception fails to accurately acknowledge the role of the technology in the experience" (Lombard & Snyder-Duch, 2001).
3. Rafaeli (1988) defines interactivity as "an expression of the extent that in a given series of communication exchanges, any third (or later) transmission (or message) is related to the degree to which previous exchanges referred to even earlier transmission" (p. 111).
4. In addition to the absence/existence of personal influence, diffusion theorists have identified other factors that may potentially influence the speed and pattern of diffusion, including: actors' involvement levels associated with the items diffused, the levels of uncertainty attached to the item, and distributions of initial beliefs related to the item within a social system (Gatignon & Robertson, 1986).
5. Critical mass "occurs at the point at which enough individuals in a system have adopted an innovation so that the innovation's further rate of adoption becomes self-sustaining" (Rogers, 2003, p. 343).

CHAPTER 2

GROUP INFORMATION EXCHANGE AND COMMUNICATION DILEMMA

> People are responding to an environment that consists of other people responding to *their* environment, which consists of people responding to an environment of people's responses.
>
> —Schelling (1978, p. 14)

A longstanding belief in marketing is that consumer word of mouth (WOM) plays a crucial role in the diffusion of information in a market (Gatignon & Robertson, 1998). Due to its stochastic nature, however, consumer WOM has been treated as a mysterious phenomenon, which is too complex to be considered as a tool for marketing communications (Rosen, 2000). The Internet, the largest communication network ever, has changed this situation. The proliferation of *virtual consumer communities* (Bickart & Schindler, 2001; Henning-Thurau, Gwinner, Walsh, & Gremler, 2004; Muniz & O'Guinn, 2001) or online feedback systems (Dellarocas, 2003) clearly demonstrates that consumer WOM is no longer a remote issue, but a key communication process, which can be monitored and even facilitated by companies. For this reason, consumer WOM in this electronically networked environment has received increasing scholarly attention for the past several years.

A virtual community is a social collectivity "sustained primarily through ongoing communication processes" (Bagozzi & Dholakia, 2002, p. 3). It is a special form of community, being neither geographically defined nor dependent on existing social relationships (Balasubramanian & Mahajan, 2001). Anyone who shares similar interests or objectives can be part of the community, regardless of social and physical boundaries. Because a virtual community is a computer-mediated aggregate of individuals distributed widely, its size can vary from a small group consisting of a few members to an extremely large community consisting of hundreds or even thousands of members. Imaginably, information can travel through the online consumer WOM networks at a remarkable speed, and have an enormous impact on the performances of products and companies.

The Internet is characterized as a *decentralized* communication environment where autonomous individual entities interact with one another (Miller, 1996; Rheingold, 1993). Since no central authorities can force or control individuals' behaviors in this environment, social collectivities like virtual communities normally *emerge* from the cooperative interactions among multiple individuals (Kollock, 1999). That is, a virtual community is a collective outcome of individuals' *volitional* communication activities, and information exchange and diffusion in this circumstance are dependent primarily on individual decisions and actions and on the interdependencies between them.

Many people have praised the Internet as a space of freedom that offers people unlimited opportunities to communicate and interact with one another (e.g., Rheingold, 1993), but they often have neglected to recognize that there is a pitfall. In this multiagent communication setting, each person may be strongly tempted to use the information provided by others without making contributions. When many individual agents interact simultaneously with one another in a group, each person is supposed to make contributions unilaterally (e.g., information provision) for all the other members (N-1 individuals) without expecting direct reciprocations from each recipient (Kollock, 1999). Because reciprocation is not required, an individual does not have to make as much of a contribution as he or she received from others, which tempts the individual to *free ride* over others' contributions.

What if everyone in the same situation decides to free ride? The initial contributors will not be reciprocated sufficiently by the others—or at all, in the worst case—and their motivations to make subsequent contributions to the group or community will fade away rapidly. If no more information is provided in the community, no one can benefit from others anymore, and the online community will eventually stop functioning. Despite the endless praises and salutations of the Internet as an ideal communication medium, numerous attempts to organize virtual communities or online forums have ended as grievous failures, and a considerable proportion of existing virtual communities are stuck in a suboptimal state. It is not difficult to find online communities or groups in which only a few dedicated players make contributions, while the rest lurk in the underground.

Without solutions to this problem, companies' efforts to organize virtual consumer forums or brand communities to interact and communicate with potential consumers will prove unsuccessful. The primary objective of this chapter is to discuss the critical issues arising from the dynamic process of multiagent information exchange, particularly in a virtual community setting. Focusing on the context of market information exchange among consumers, this chapter is aimed primarily at understanding the group information exchange process and identifying the network-structural as well as individual psychological factors encouraging or hindering consumers' electronic word-of-mouth (eWOM) motivations and behaviors.

Social Exchange and Diffusion of Information

Consumer WOM is a social exchange behavior. Despite the apparent connection between the social-exchange theories and the studies of the diffusion process, the two lines of inquiry have rarely converged. Social-exchange theorists have attempted to understand the process of *resource exchange* among individuals (Cook & Whitmeyer, 1992), but have paid little attention to how the social-exchange process in a small-group situation is translated into large-scale diffusion. On the contrary, diffusion theorists have focused on the various determinants of the large-scale diffusion of innovation, including the personal attributes of communicators as well as social networks (Mahajan, Muller, & Bass, 1990; Rogers, 2003; Valente, 1995). Little attention has, however, been devoted to explicating the microprocess of information exchange among multiple individuals in this domain.

For the past several decades, most diffusion-related studies have concentrated on three general subjects: (1) identifying *market influencers*, such as innovators (e.g., Engel, Blackwell, & Kegerreis, 1969; Engel, Kegerreis, & Blackwell, 1969; Midgley, 1976; Midgley & Dowling, 1978), opinion leaders (e.g., Chan & Misra, 1990; Myers & Robertson, 1972; Richins & Root-Shaffer, 1988; Robertson & Myers, 1969), or market mavens (e.g., Elliott & Warfield, 1993; Feick & Price, 1987; Wiedmann, Walsh, & Mitchell, 2001; Williams & Slama, 1995); (2) revealing the *referral networks* of consumers (e.g., Brown & Reingen, 1987; Reingen & Kernan, 1986); and (3) investigating the effects of word of mouth on consumer decision making (e.g., Herr, Kardes, & Kim, 1991) and purchase behavior (e.g., Arndt, 1967). In this line of research, a social network has been postulated as a "static" structure that was given prior to the interactions among people, which functions merely as a route for information flow. Thus, the information exchanges among multiagents (N-person exchange) and those occurring between two persons (dyadic exchange) are not differentiated, because the N-person exchange is conceived as an additive sum of multiple dyadic parts: a large-scale diffusion is understood as an aggregate of the events of dyadic information transmission.

However, sociologists have long recognized that N-person interaction is different from a dyadic interaction in terms of the actors' views toward the situation. Simmel (1950), in his famous article, pointed out that the existence of a third actor, as in a *triadic* relationship, might fundamentally alter the actors' perceptions of the social interaction situation. Within a triad, each actor should seriously consider the relationships between the other two: whether the other two are close to each other, communicate frequently, and/or agree on certain issues that may influence one's attitude, choice, and behavior (Coleman, 1988). Diffusion theorists have long postulated that information diffusion occurs through social networks consisting of a multitude of individuals, but they have not paid attention to how the multiagent communication settings affect individual actors' communication behaviors.

For example, Frenzen and Nakamoto (1993) examined how the strength of the social relationship between an information sender and a potential recipient influences the likelihood of the information giver's *opportunistic* behavior (hoarding information). An assumption underlying their study is that if each dyadic information transmission is successful, the information will be diffused through the network. This situation, however, may not apply to the *many-to-many* simultaneous interaction situations, as in virtual communities. As mentioned before, when multiple individuals engage simultaneously in information exchange, each individual considers not the action of a particular individual but the aggregate activities of N-1 individuals. If the others communicate actively, the person may feel that his or her further contribution is unnecessary, and such moral hazard or opportunistic behavior may make it difficult to achieve a collective goal.

Figure 2.1 compares two different models of information exchange. Previous diffusion research has postulated consumer

FIGURE 2.1. Sequential vs. simultaneous information exchange.

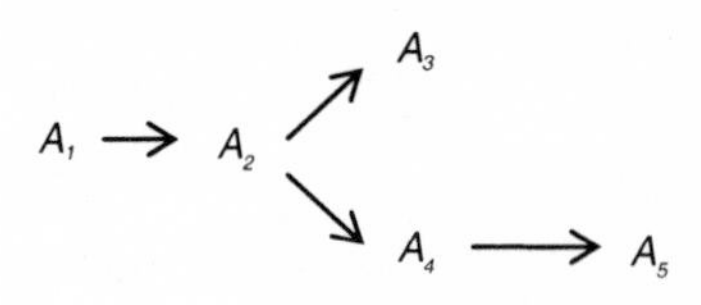

(a) Sequential Transmission

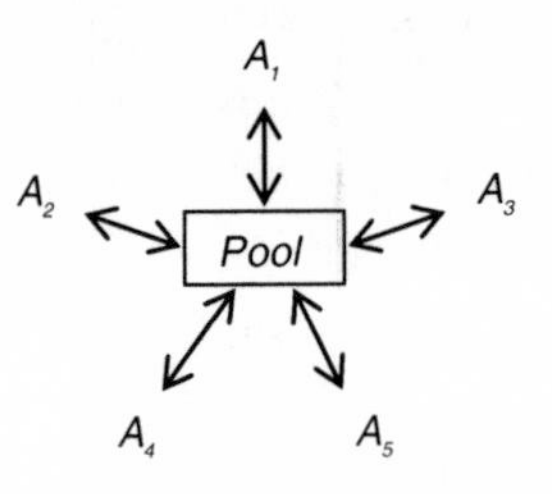

(b) Simultaneous Exchang

WOM as the process of *sequential* information transmission. In this model, what each actor should consider is only the relationship with a few direct recipient(s), not with all recipients in a network. Social exchange theorists, on the other hand, have been concerned more with the model of *simultaneous* information exchange. In this model, each actor is supposed to know the existence of multiple other actors in a group and should consider their behaviors in deciding his or her prospective courses of action.

Distinguishing between the two models is very important because they create quite different environments for consumer WOM. The sequential communication model illustrates the traditional offline WOM environment, in which consumers relay information and opinions through the chains of social connections, and the number of recipients whom each individual can reach is quite limited. On the other hand, examples of the simultaneous information exchange model can be found in the group information-exchange situations like virtual communities or online feedback systems (e.g., eBay). Because the uniqueness of the Internet comes from its facilitating multiagent information exchange, understanding consumer's electronic word-of-mouth (eWOM) behaviors requires grasping how individuals' communicative motivations are shaped and conditioned by the various characteristics of simultaneous communication settings. In order to illuminate this interaction process among many actors, in the following sections, *social exchange theories* (e.g., Emerson, 1976) and the frameworks developed in game theory are discussed in detail.

Narrow Rationality and Social Dilemma

Adam Smith, the father of modern economics, said that a market economy works most efficiently when each individual pursues his or her own selfish interests. Symbolized by the "invisible hand," the gist of his idea is that individuals' pursuit of personal interests leads to the good for everyone in a society. Egoistic motivations of individuals are not in conflict with the collective interest of a market or even a whole society. For the past several decades, however, social scientists, including economists, sociologists, and psychologists, have found many situations that violate this basic principle of economics. These peculiar situations, in which individually rational decisions lead to collectively irrational consequences, have been called "social dilemmas" (Hardin, 1982).

Social dilemma is defined as a *deficient equilibrium* between individual and collective interests. "It is deficient in that there is at least one other outcome in which everyone is better off. It is an equilibrium in that no one has an incentive to change their behavior" (Kollock, 1998, p. 184). For example, keeping an environment clean is beneficial for everyone living in the same district, but individuals may be tempted not to make their own contributions (e.g., removing trash on streets) and simply to benefit from others' contributions. If everyone cooperated to make their environment clean, they would surely be better off, but the reality often deviates from this ideal situation.

Another example is a public library. Since people are allowed to use a public library regardless of whether they make contributions (e.g., donating money) to it, most of them do not have any incentive to make monetary contributions to it. As a result, such a public library can hardly be maintained without governmental subsidy or other external funds. As the aforementioned examples imply, social dilemma has a particular incentive structure, in which the best outcome is obtained when everybody cooperates—but individuals have few incentives to do so. In a similar vein, Komorita and Parks (1996) defined a social dilemma as "a situation in which a group of N-persons ($N \geq 2$) must choose between maximizing selfish interests and maximizing collective interests" (p. 8).

A social dilemma occurs when the group interactions or communications among people involves *public goods*. A public good refers to a type of good defined by two characteristics: (1) *Jointness of supply* and (2) *nonexcludability* (Hardin, 1982; Olson, 1965). If a product is in joint supply, one's consumption of the product does not decrease the proportion that others can consume. For example, a person's watching a television program aired by a public broadcasting station (PBS) does not affect the opportunity for others to enjoy the program. Secondly, nonexcludability means that once a public good is provided, no one can be alienated from its benefit—which means that no one can exclusively benefit from a public good. In contrast, a private good (e.g., automobile, food) is consumed only by its owner, unless the owner wants to share it with others.

According to Hardin's (1982) formal expression of the difference between public and private goods, in a group consisting of N actors, the total consumption (X) of a private good can be expressed as an additive sum of individual consumption (χ_i): $X = \Sigma_{i=1}^{n} x_i$. For a public good, however, an individual *i*'s consumption of the good is the same as an individual *j*'s consumption of it, and so forth: $\chi_1 = \chi_2 = \chi_3 = \ldots \chi_{n-1} = \chi_n \ldots$. This does not mean that the amount of consumption of a given public good should be equal among individuals, but it implies merely that a person's consumption of a public good does not influence others' consumption of that good. This formal representation, however, applies only to a public good dilemma, not to a "social trap" situation in which one's consumption of a good diminishes the proportion of others' consumption.[1]

Olson's (1965) classic formal expression of a social dilemma (public good dilemma) is based on three simple terms, as follows: $A_i = V_i - C_i$, where V_i (= individual *i*'s gross benefit of making a contribution), C_i (= individual i's cost to make a contribution), and A_i (= individual i's "net" benefit of making a contribution). If $A_i > 0$, individuals will be motivated to cooperate with others, and the pubic is likely to be provided for successfully. On the contrary, if $A_i < 0$, individuals may be reluctant to make contributions, even though they know that providing the public good is beneficial for everyone.

According to Olson's reasoning, public goods cannot be provided in most instances in which individuals behave rationally, because A_i can hardly be positive in a short term period. In other words, an individual can get a positive return instantly from free riding, but the outcome of cooperation is either minimal or comes later. Furthermore, he insisted that this public good dilemma will become worse as the size of a group increases because the fractional benefit of an individual i's contribution (A_i) becomes smaller and smaller as the group gets bigger and bigger, and because it gets more difficult to make a discernable contribution to a group. Based on this reasoning, he concludes that a large group will fail, while a small group is likely to succeed.

Public goods, however, do not constitute a sufficient cause of a dilemmatic situation. There are two additional conditions to be met for a social dilemma to occur. Assume that an individual actor i has two alternatives of action, either cooperation, denoted C_i, or defection, denoted D_i. First, the payoff of an individual i's defection (D_i) must be larger than that of cooperation (C_i): $D_i > C_i$. If C_i is bigger than D_i, there is no reason to defect and harm the collective interest. Second, the outcome when everyone cooperates (C_a) must be larger than the outcome when everyone defects (D_a), which means that $C_a > D_a$. If cooperation yields nothing more than when everyone pursues selfish goals, there should be no incentive for individuals to cooperate with others. Figure 2.2 illustrates two hypothetical situations in an N-person exchange setting. The value in each cell indicates only the outcome for an individual's choice (C or D) that corresponds to the choice of N-1 others.

Figure 2.2 (a) is the N-person prisoner's dilemma (NPD), in which the order of payoffs follows *DC* > *CC* > *DD* > *CD*. In this case, the incentive for defection is greater than incentive for cooperation because one can get a positive payoff (= 8) even if the others also defect. In contrast, cooperation is not a good choice, because it would not yield the best outcome for all and may lead to the worst payoff (= –8) for the person if the others make no contribution. On the other hand, Figure 2.2 (b) illustrates a situation of "assurance game" (AG), in which cooperation leads to the best outcome (*CC* > *DC* > *DD* > *CD*). In this situation, the only requirement for cooperation is *trust* in others (Kollock, 1998). If one can be sure that others would cooperate, then cooperation must be the best choice and lead to *Pareto optimality*.[2] Otherwise, defection may be a safer choice, though it would lead to a deficient equilibrium.

Many social dilemmas occurring in the real world can be categorized into either of the two different types of N-person dilemmas illustrated previously. In an AG situation, people may be more easily motivated to cooperate because cooperation leads

FIGURE 2.2. Multiple-person dilemmas.

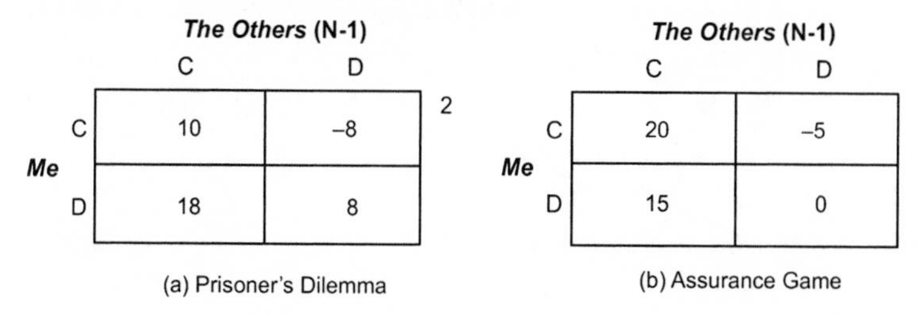

to the best outcome for everyone in a group. Thus, individuals in this situation can be made cooperative if they are shown evidence that others are also cooperative. Social dilemmas falling in the category of the NPD situation, however, are more problematic because defection dominates cooperation. In this situation, simply showing others' cooperative behaviors may not be helpful for increasing cooperation, but instead increases the degree of temptation to defect. For this reason, social dilemma researchers have suggested that increasing cooperation in the NPD cases requires transforming its incentive structure from that of NPD to that of AG (Kollock, 1998). For example, it has been suggested that the order of outcomes in NPD ($DC > CC$) can be reversed to that in AG ($CC > DC$) by introducing a *sanctioning system* (Yamagishi, 1988) for rewarding cooperation and punishing defection.

COMMUNICATION DILEMMAS IN A VIRTUAL COMMUNITY

The crisis of a virtual community using a publicly shared communication channel or space, such as an electronic bulletin board, comes from a situation where "it is in the collective interest of network members to communicate, but in each separate interests to hoard information" (Bonacich, 1990, p. 448). Information is a public good in the sense that it satisfies the two conditions—*nonexcludability* and *jointness of supply*. If a person posts a message containing useful information on an e-board, this information can be shared with anyone who can access the board. Also, one's consumption of the information does not decrease the opportunity for others to consume it. This means that anyone can exploit information provided by others without making his or her own contributions, and the negative effects of this exploitative behavior on the group may not be revealed immediately.

In addition to the characteristics of information as a public good, there are some other conditions that make a virtual community particularly vulnerable to the free rider problem. First, when more than two individuals ($N \geq 3$) interact in a group, the effects of one's defective behavior are diffused over N-1 individuals, while in a two-person case, one's defection directly harms the other (Dawes, 1980). This means that in the N-person case, a person may have much less social responsibility for cooperation than in the two-person case. Yamagishi and Cook (1993) call this *the effects of responsibility diffusion*.

Second, in the N-person interaction situation, the effects of each individual's contribution on the group or community should be smaller compared to the two-person case. In this situation, individuals may not be motivated to contribute because "one small contribution would hardly make a difference" (Komorita & Parks, 1996, p. 55). That is, people's *self-efficacy* (Bandura, 1986) of contribution will be lowered as the group size increases, as Olson (1965) pointed out earlier. Finally, if the exchange is iterated over time, it would be very difficult to identify and directly control anonymous noncooperators in the N-person case, while the noncooperator in the two-person case can be controlled directly by punishment or reward in the next stage (Dawes, 1980).

Bonacich (1990) named this particular dilemmatic situation arising from a group communication as a "communication dilemma." Providing information on an e-board requires a certain amount of *behavioral costs*[3] (Robben & Verhallen, 1994; Verhallen & Pieters, 1984). If paying the cost instantly produces extra benefit, people will be motivated to provide information. If not, they will not do so. In a typical virtual community, however, one's contribution is neither enforced nor reciprocated directly by those who benefited from it. Under this circumstance, an individual's contribution becomes *unilateral* rather than reciprocal, and one's unilateral contribution is "an invitation to exploitation" (Takahashi, 2000, p. 1107). This form of unilateral exchange, based on a *social system* involving three or more individuals "in which each actor gives to another but receives from someone other than to whom he gave" (Bagozzi, 1975, p. 33), is called a "generalized exchange" (Ekeh, 1974; Levi-Strauss, 1969). Social dilemmas result from the incentive structure of this particular form of exchange (Yamagishi & Cook, 1993).

Human social exchange can be grouped into two basic forms: *restricted* vs. *generalized* exchange. Figure 2.3 illustrates these two forms of exchange. *Restricted*, or *balanced* (Sahlins, 1974), *exchange* refers to a type of *reciprocal* exchange between two individual entities. Gift exchange between two friends is an example of this type of exchange (Pieters & Robben, 1998). In this exchange form, each person is socially bound to reciprocate the favors of the other, because ignoring one's favor may harm the social relationship.

Generalized exchange, on the other hand, can be regarded as a form of exchange in which "each actor provides resources

FIGURE 2.3. Basic forms of exchange.

A_1 ⇄ A_2

(a) Restricted Exchange

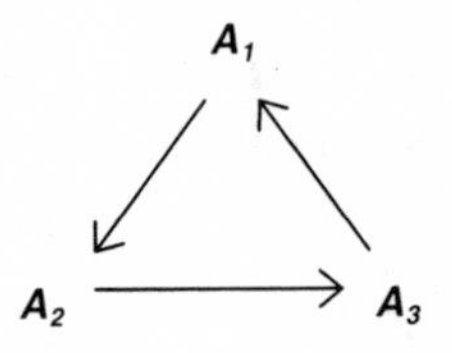

(b) Generalized Exchange

at some time and eventually receives some benefit in return" (Takahashi, 2000, p. 1107). Helping a stranded driver on the road or giving useful information to others are examples of generalized exchange. As shown in the Figure 2.3 (b), A_1 is supposed to receive something from A_3, not from A_2, who will benefit from the contribution of A_1. This means that whether A_1 benefits from A_3 is not dependent on A_1's contribution to A_2. In other words, the probability that A_1 will benefit from A_3 is separate from his or her relations with A_2, and A_1 may be strongly tempted not to make a contribution to A_2.[4] For this reason, individuals' cooperative motivation may be weakened because they cannot be sure whether others will also be cooperative.

Generalized exchange is a cornerstone of building up a social group or community,[5] and, at the same time, is the major cause of social dilemmas. This seemingly discrepant nature of generalized exchange has led scholars to many perplexing questions. First of all, how can individuals be cooperative in a society at all, if a generalized exchange is so vulnerable to the free rider problem? Why do people make contributions for others without expecting immediate reciprocity? Simply attributing the prosocial behavior to *altruistic motivation* (e.g., Hemetsberger, 2002; Sahlins, 1974) or *collective intention* (e.g., Bagozzi & Dholakia, 2002) gives rise to another question: Why does such a motive exist? Why do people sometimes care about the well-being of others, and not care at other times?

A breakthrough on this subject has emerged from the pioneering works in social-exchange/collective-behavior research to investigate how cooperation among adaptive individuals is formed. In the past few decades, social scientists have found that an individual's cooperative behavior is an outcome of his or her *adaptive* response to the immediate social environment consisting of others (Axelrod, 1984; Schelling, 1978). The crucial notion found by these studies is that a group-level outcome, like the cooperation among individuals, can emerge from the relatively simple decision rules guiding individual behaviors—*maximizing rewards, minimizing costs* (Macy & Willer, 2001): complex aggregate outcomes do not necessarily arise from complex individual behaviors. Although true "altruism" undoubtedly exists (Piliavin & Charng, 1990), altruistic motivations may not be always necessary for the emergence of cooperation.

"Adaptation" in this context means that one's resource giving results from his or her consideration of the benefit or loss defined by a given exchange situation. It should be noted that adaptation is not the same as the utility-maximization principle. Utility (or expected utility) maximization in economics is a normative decision-making principle that is assumed not to be influenced by the varying choice contexts (von Neumann & Morgenstern, 1947). The concept of adaptation implies that a person's decision making and behavior should be understood as a flexible process, depending on the characteristics of the decision contexts that are given (Bettman, Luce, & Payne, 1998). This means that an individuals' choice of cooperation or defection is a context-dependent behavior, influenced by the incentive structure of a situation.

Stimulated by this notion, contemporary social scientists have conducted a series of intriguing experimental studies by focusing on the structures emerging from a combination of actors and relations among them. The topics covered by those studies include social dilemmas (e.g., Bonacich, 1990; Yamagishi & Cook, 1993), power relations in exchange (e.g., Silver, Cohen, & Crutchfield, 1994; Skvoretz & Lovaglia, 1995; Skvoretz & Willer, 1991), and the effects of exchange structures on actors' emotions (e.g., Lawler & Yoon, 1996, 1998), among many others. These pioneering studies have repeatedly confirmed the notion that one's cooperative or defective motivation and behavior are *environmentally defined* to a considerable extent. If the immediate social environment consisting of others is changed, an actor's motivation and behavior may change correspondingly. This notion pinpoints the idea that studying the characteristics of group communication in which individuals interact with one another may provide crucial knowledge for understanding and even facilitating cooperative communication among consumers.

From Group to Network: A Structural Solution

Social dilemma researchers have proposed a variety of solutions to the free-rider problem, which are generally categorized into two types of strategy (Kollock, 1998; Messick & Brewer, 1983). The first is an *individual psychological solution* that is aimed at changing the ways people perceive the situation of social dilemmas (e.g., Rosen & Haaga, 1998). Communication campaigns to persuade people to contribute to conserving the natural environment is an example of this approach. The second is called a *structural solution*, which is aimed at transforming the incentive structure of a given situation. This approach can be distinguished further into categories of *direct* and *indirect* intervention (van Vugt, 1997). Direct intervention refers to the strategy to change the given incentive structure by directly providing a sanctioning system. It has, however, been pointed out that these direct interventions may cause another dilemmatic situation, which is called a "second-order social dilemma"[6] (Yamagishi, 1988). On the other hand, the indirect intervention refers to the ways of arranging the given incentive structure on a long-term basis by altering the structure of exchange (e.g., Molm, 1994; Yamagishi & Cook, 1993).

Among many suggested solutions to the social dilemmas, this research focuses primarily on the indirect structural approach as a possible solution to communication dilemmas in virtual communities. More specifically, the focus is on the strategy to transform a public communication space into *a network of multiple private spaces*. The two key elements of this structural solution are (1) *privatization* and (2) *networking*. First, privatizing a public good—dividing a common property into private segments—has been considered as a possible solution to the free-rider problem, because people tend to care better for their own properties than for a public one (Ostrom, 1990). In reality, transforming a public corporation into a private one has often been found to be an effective way to make it operate more efficiently and competitively (van Vugt, 1997).

This privatization, however, may not be the sole solution to the communication dilemmas, because simply dividing a public communication space into separate private ones makes it difficult for individuals to communicate for a collective interest. As a result, the privatized spaces should be interconnected to form a *communication network*.

Even though both a group based on a public channel and a network of privatized segments are collective entities, their internal structures for communication are radically different from each other. Figure 2.4 illustrates this structural solution graphically. Yamagishi and Cook (1993) called the first a *group-generalized exchange*, in which "group members pool their resources and then receive the benefits that are generated by pooling" (p. 237). Exchanging information through an electronic bulletin board, which is typical in most virtual communities, is an example of this structure. The second is called *network-generalized exchange*, in which actors exchange benefits through interpersonal networks, not through a public resource pool.

FIGURE 2.4. Group vs. network-generalized exchange structures.

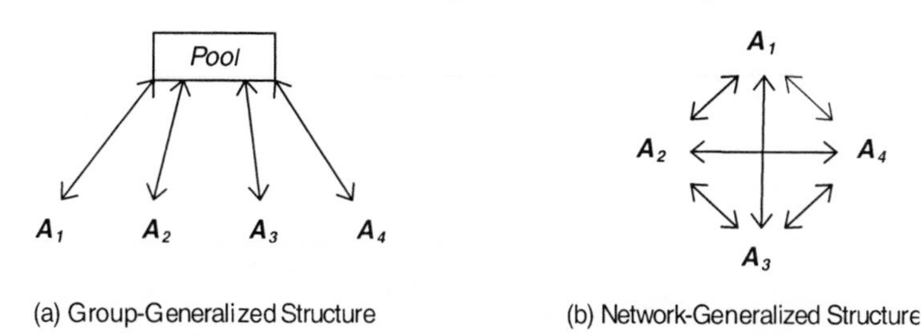

(a) Group-Generalized Structure (b) Network-Generalized Structure

WOM communication among individuals, which is based on the social connections among them, may be regarded as an example of this exchange structure. For simplicity, let us denote the two structures as G_{EX} (group generalized) and N_{EX} (network generalized), respectively, in the remaining discussion.

The major difference between the two structures is that G_{EX} is based on the existence of a *public resource pool*, while N_{EX} is sustained by the *exchange networks* among actors. Through a series of experiments, Yamagishi and Cook (1993) found that people in the N_{EX} condition tended to be more cooperative (giving resources to others) than those in the G_{EX} condition.[7] Why does such a difference in cooperation exist? First of all, Yamagishi and Cook point out that G_{EX} is disadvantageous in the sense that the responsibility of maintaining the public resource pool is diffused evenly over all members, which allows one to free ride on others' responsible behaviors. This is referred to as "the diffusion of responsibility effect." This means that each individual in G_{EX} can benefit from the N-person information exchange, with a very low cost.

Let us assume, hypothetically, that there is a minimum level of costs required for providing a public good (e.g., maintaining an online community status quo), which is denoted as C_0. In

G_{EX}, C_0 is an additive sum of the cost allocated to an individual actor *i*, which is denoted c_i, thus, $C_0 = \Sigma^n_{i=1} c_i$. This means that the cost allocated to an individual becomes $c_i = C_0/N$, which indicates that c_i decreases as N increases. Furthermore, G_{EX} allows the cost allocated to an individual to be paid by the contributions made by others, denoted as *K*. If $K \geq C_0$, individuals' motivation to pay the costs will drop rapidly, because paying additional costs does not make any noticeable difference in the provision of a public good. This is called *order effects*, which refers to the tendency for people to free ride on the initial contributions made (Heckathorn, 1996).

On the other hand, if the public information pool is segmented into private communication spaces, each private space reveals both the owner's defective and cooperative behaviors. In other words, whereas G_{EX} allows individuals to exploit public goods provided by others, N_{EX} allocates the cost for a public good to those who benefit from them, which is termed "internalization of externalities" (Coleman, 1988). In N_{EX}, each individual actor *i* should pay the minimum cost for maintaining his or her own private segment (c_i), and this cost does not decrease as the group size increases. Because c_i does not vary depending on the others' contributions (*K*), individuals' exploitative motivation may be blocked by the cost incurred. In other words, if someone does not frequently update the contents of his or her own space in a network, the suboptimal state of the private space will be easily revealed to other network members, and the owner may receive negative feedbacks from them.

Personal responsibility, however, is not the end of the story. Collective-behavior researchers have found that an individual's cooperation is also dependent on the individual's perceived *efficacy of contribution* as well. If one's contribution cannot make a noticeable difference and is a waste of time, individuals will not cooperate, even though they believe it necessary (Gould, 1993). In terms of the efficacy of contribution, G_{EX} also has a disadvantage, in that all the contributions made are mixed into an identical space, in which no one's contribution can stand out. If others already made sufficient contributions, and one's additional contribution is not recognizable, then one's cooperative motivation will be substantially reduced. Conceivably, this low efficacy situation in G_{EX} may be exacerbated when the group size gets bigger (Olson, 1965).

In contrast to G_{EX}, N_{EX} as a network of private segments increases the visibility of one's cooperation or defection. The recent popularization of the new system of online communication called *blogs* (Web logs)[8] can be regarded as an excellent example of N_{EX}-based communication. In this system, individuals communicate with others based on *networks* of private communication spaces, not on a publicly shared space. If an individual actor posts information or a message on his or her own page, others who have connections with the original information provider are notified of the information update, and then are able to easily access the information.

ENDNOTES

1. Social dilemmas are categorized into two types. The first is called a "public good dilemma," which refers to a situation in which providing a public good (e.g., environmental conservation) is made difficult by individuals' selfish behaviors. The second is called a "commons dilemma" or "social trap," referring to a situation in which a commonly shared property (e.g., fish in a pond) is made obsolete by the owners. Both types of social dilemmas are a result of the nonexcludability of a public good, but the commons dilemma is additionally related to the subtractability of benefits. In a public good dilemma, one's use of the public good does not reduce others' ability to benefit from it. However, in a commons dilemma, one's use of a public property reduces how much others can use. Komorita and Parks (1996) pointed out that a public good dilemma is associated with short-term negative and long-term positive outcomes, while a social trap is, on the contrary, associated with short-term positive and long-term negative outcomes.
2. Pareto optimality is defined as the outcome that disadvantages no one involved.
3. Behavioral cost is defined as the ratio between the behavioral price and the behavioral budget of an individual. Verhallen and Pieters (1984) broke down one's behavioral costs into the following elements: time price (T) and budget (TB), psychological price (Ps) and budget (PsB), and physical price (Ph) and budget (PhB). Given these elements, behavioral cost (BC) is formally defined as follows: $BC = \sum (T/TB, Ps/PsB, Ph/PhB)$.
4. An assumption in this figure is that each actor does not know the relationship between the other two, because if A2 knows that A1 is the recipient of A3's contribution, he or she can punish A1's non-cooperation by making A3 stop contributing in the next stage (Yamagishi & Cook, 1993). In other words, actors in this group are assumed to be unaware that they are in a chained relationship. This chained relationship is just one of many possible forms of generalized exchange.

Chapter 3

Hypotheses

The Effects of Structural Change

Summarizing the discussions in the previous chapter, N_{EX}-based group communication may be advantageous with respect to the two criteria—*social responsibility* and *contribution efficacy*. First, failure to provide sufficient information in N_{EX} may be detected easily by others, and will return negative consequences such as bad evaluations to the owner of the site. If this situation goes on, the private site, which contains significantly less information than others in an information exchange network, may no longer attract visitors, and the value of the site as well as the owner's reputation will be significantly lowered. In contrast, an individual's non-contribution may be less conspicuous in G_{EX} than N_{EX}.

Second, in the system of N_{EX}, one's contribution may be displayed more prominently than in G_{EX}. Posting information in a privately held rather than in a public place may stimulate individuals' communicative motivation to provide information and contents that others can see, whereby they can obtain *social approval* from others more efficiently. In the *blog* system, for example, a person can organize freely his or her contributions (e.g., information, essays) to make some of them more visible to visitors. Furthermore, the contributions made are accumulated in the private space, which can be shown to visitors anytime, unless the owner intentionally erases them. Applying this notion to the context of consumer information exchange, it may be reasonably inferred that individuals in N_{EX} may be considerably more motivated to communicate with others in N_{EX} than in G_{EX}. Therefore:

H1. *Consumers will make substantially more contributions when they communicate with one another in N_{EX} than in G_{EX}.*

In addition to the overall increase in cooperation by the structural change, it may also be hypothesized that N_{EX} motivates individual consumers to engage in communicating with others in a more stable manner than does G_{EX}. Because each individual is supposed to maintain his or her own site in N_{EX}, this situation may force them to update their own sites on a regular basis. In contrast, consumers in G_{EX} may behave *passively*, depending on other members' activities because their degree of personal responsibility and contribution efficacy is inseparably related to the how much others contribute. This means that individuals in G_{EX} may become less and less cooperative, as the others in the group make more contributions. Since everyone waits for others' contributions in G_{EX}, relatively few original contributions will be made, and the cooperation will decrease substantially as time goes by. Contrastingly, an individual actor's cooperative motivation may be less influenced by the level of others' contributions in N_{EX}, and the overall contribution level may be relatively more stabilized than in G_{EX}.

H2. *The level of cooperation among individuals will be more stable when they communicate with one another in* N_{EX} *than in* G_{EX}.

The Effects of Group Size

The effects of group size on cooperation and collective behavior have long been a central concern in social dilemma research since Olson (1965) argued that a large group should be eventually doomed to fail in terms of a provision of public goods. In his argument, a large-sized group is less efficient in motivating actors to contribute, because individual contribution becomes unnoticeable, and the fractional benefit from making a contribution is easily exceeded by the cost of doing so. Consistent with his notion, scholars studying collective behaviors have often found that it becomes more difficult to motivate individuals to cooperate with others as the size of a group increases (Komorita & Parks, 1996).

For example, Kerr (1989) showed that the failure of large group resulted primarily from the negative effects of group size on individuals' efficacy of contribution. He showed in a formal way the relationships between an actor's self-efficacy (λ) and the size of a group as follows:

$$\lambda = \binom{N-1}{p-1} P(C)^{p-1} [1 - P(C)]^{N-p}$$

Where N is the size of a group, p is the minimum number of contributors needed for providing a public good, and $P(C)$ is the probability of others' contribution. As indicated in the formal expression, an individual actor's perceived efficacy of contribution (λ) will increase as *P*(*C*) increases, and *p* decreases. On the other hand, λ will decrease as N increases, if the other conditions are held constant. In a series of experiments, he found empirical evidence supporting the formalized relationship between self-efficacy of contribution and group size. Since people hate for their valuable resources (e.g., time) to be wasted by making a contribution yielding negligible outcomes, it seems rational that they are strongly tempted not to cooperate in a large-group situation.

The deteriorating effects of group size, however, may not apply to all situations. What should be noted is that previous arguments of the group-size effects are based on the conception of a group as mere aggregate of independent individuals, which does not have any prespecified structure of communication. That is to say, prior studies have considered only the generalized exchange through a public resource pool, which is G_{EX}, disregarding the possible alternative exchange structures that may be resistant to the effects of group size. G_{EX} appears to be particularly vulnerable to the effects of group size because an individual actor's contribution efficacy and social responsibility are minimized as the group gets larger. Contrastingly, N_{EX} is a network of privatized spaces, in which each person's responsibility for making contributions may not change much as the size of the group increases, because the visibility of one's defective behaviors is not influenced by the amount of others' contributions in N_{EX}. Also, an individual's self-efficacy for making contributions in N_{EX} may not be negatively associated with the size of a group, but can increase as the group gets bigger because the increase of group size provides more and more potential audiences for his or her communication activities.

This implies that networking individuals can be a more efficient way to organize the *many-to-many* communication than merely grouping them, especially when a large number of individuals are involved. In other words, a network can be conceived of as a robust form of communication involving many people, compared with a group based on a public resource pool. Based on the discussions, it may be inferred that individuals' cooperation levels in N_{EX} may be influence positively by the size of a group, or at least, it may not be affected negatively, while those in G_{EX} are negatively affected by the group size. Therefore:

H3. *Consumers' level of cooperation in information exchange is likely to increase or change minimally in* N_{EX}*, but is likely to be lowered in* G_{EX} *as the group size increases.*

Social Value Orientation

Even though N_{EX} may be superior to G_{EX} in motivating individuals' social cooperation, it remains unclear what such structural effects will produce. Therefore, it may be instructive to find out which of the two criteria discussed previously—personal responsibility and contribution efficacy—differentiates N_{EX} from G_{EX}. Does N_{EX} motivate individuals to cooperate by making them sensitive to personal responsibility for managing their own spaces, by providing opportunities for them to make contributions efficaciously, or both? This question is related to the psychological motives underlying people's responses to the structural change from G_{EX} to N_{EX}. Conceivably, individuals may respond differentially rather than uniformly to the structural change, depending on their personal characteristics: some people may change their behaviors more or less radically than others.

In order to answer the question, this study considers the concept of *social value orientation* (Messick & McClintock, 1968), which is defined as "stable preferences for certain patterns of outcomes for oneself and others" (van Lange, de Bruin,

Otten, & Joireman, 1997, p. 733). That is, social value orientation refers to an individual's psychological propensity to overestimate (or underestimate) the importance of outcomes for others compared with those for oneself. The implication of this concept is that individuals' cooperation or noncooperation in social contexts may be partly determined by their personality characteristics. In the social value orientation research, individuals are usually distinguished into three categories—*prosocials*, *individualists*, and *competitors*. Individualists and competitors are often combined to form a category named *proselfs* (de Cremer & van Vugt, 1999).

Prosocials is a category of individuals who tend "to maximize the outcomes for both themselves and others (i.e., cooperation) and to minimize differences between outcomes for themselves and others (i.e., equality)" (van Lange et al., 1997, p. 733) by focusing on the joint outcomes. In contrast, *individualists* have a tendency to maximize their own outcomes while disregarding the outcomes for others, and competitors try to maximize the difference between their own outcomes and others' outcomes. *Proselfs*, a combination of the individualist and competitor categories, can be regarded as a category of individuals who are more concerned with the outcomes for themselves than for others. Applying this concept to the social dilemma situation, researchers have repeatedly found that *prosocials* behave more cooperatively than *proselfs* in dilemmatic situations (e.g., de Cremer & van Vugt, 1999; Kramer, McClintock, & Messick, 1986; Utz, 2004; van Vugt, 1997).

If N_{EX} increases people's cooperation by making them more sensitive to personal responsibility, it is expected that the cooperative motivation of *prosocials* may not be differentiated much by the structural change, because they are inherently motivated to cooperate with others. Thus, imposing more responsibility on *prosocials* would not encourage *them* to cooperate more. Rather, imposing such external social pressures may compel *proselfs* to cooperate because they may respond sensitively to the potential negative consequences of noncooperation for themselves.

On the contrary, if N_{EX} enhances people's contribution efficacy, *prosocials* may respond sensitively to the structural change because it may serve as an extra incentive encouraging them to cooperate more for increasing the joint outcomes for self and the group as a whole. In this situation, conversely, *proselfs*' cooperation level may not change depending on the structural change, as they are inherently lacking in motivation to cooperate for achieving a group goal. Let us name the first explanation the *responsibility-effects hypothesis* (H3a), and the second one the *efficacy-effects hypothesis* (H3b). The hypotheses can be stated as follows:

H4a. *N_{EX} will motivate proselfs to cooperate more than will G_{EX}, while the level of cooperation of prosocials will not differ substantially between the two structures (responsibility-effects hypothesis).*

H4b. *N_{EX} will motivate prosocials to cooperate more than will G_{EX}, while the level of cooperation of proselfs will not differ substantially between the two structures (efficacy-effects hypothesis).*

Formation of Group Identity

Social scientists have long believed that it is necessary to develop *group identity* (perceived membership of a group) or *group intention* in order to strengthen the solidarity among individuals as a group (Messick & Brewer, 1983). Studies in social psychology have constantly confirmed that simply categorizing individuals as a group may increase their favoritism of in-group members against out-group members, which, in turn, induces them to act cooperatively for achieving group goals (e.g., de Cremer & van Vugt, 1999; Tajfel, 1978; Wit & Wilke, 1992). Social psychologists have attributed these "social categorization effects" to people's psychological tendency of drawing coherent *self-identities* by identifying with a group (Tajfel, 1978). Thus, it has been believed that the cooperation among individuals may become fragile, particularly in a dilemmatic situation without a solid group identity commonly shared.

On the Internet, however, it becomes very difficult to develop such a group identity, for two reasons. First, identifying oneself as part of a group is facilitated when the group's goal and boundary are so clear that in-group members are easily distinguishable from people outside the group (Tajfel, 1978). Since the Internet usually involves numerous *anonymous* individuals with different interests and purposes, the distinction between members of the group and nonmembers is not always clear, which obscures the goal of the group or community. Second, the interactions among individuals on the Internet tend to be more *temporal* than continuous. The interactions among consumers on the Internet are made usually for specific purposes like searching for information, and there remain no reasons for them to interact further once the needs are satisfied. Since a solid group identity is developed through lasting interactions among actors, this temporal nature of online interactions may make it difficult to develop group solidarity.

Consistent with this inference, communication researchers have repeatedly found empirical evidence that CMC is ineffective for developing social relationships because of its lack of available *social cues* (Walther, 1996). For example, Sproull and Kiesler (1986) found that CMC tended to reduce social context cues, which, in turn, deregulate communication between people.

Also, Sussman and Sproull (1999) found that CMC made communication between people impersonal, which made individuals less resistant to telling bad news to others. Further, some researchers have reported that the Internet reduces the psychological well-being of people by isolating them from one another (Kraut et al., 1998).

Let us return to our comparison of G_{EX} and N_{EX}. N_{EX} is expected to increase people's cooperative motivation by segmenting a public communication space into private ones. However, one may question whether we can still call the aggregate of individuals who manage only their own private properties a *group*. Further, it may be argued that breaking down a group, which shares an information pool among multiple private properties, may harm the identity of the group because cooperation in the structure no longer originates from group identification but from the desire to increase personal rewards while avoiding losses. This potential conflict between group identity and privatization strategy implies that changing G_{EX} to N_{EX} may yield cooperative outcomes temporarily, but eventually harm the solidarity of a group. In other words, the increase in cooperation in N_{EX} may be due primarily to the increased sensitivity to the selfish interests of individuals, which eventually harms the formation of a common identity as a group.

Privatizing the public information pool, however, may not necessarily make individuals egoists. Instead, changing G_{EX} to N_{EX} may make group cooperation less dependent on the perceived identity of a group because the activation of one's *self* does not always stimulate his or her orientation to selfish benefits. For example, Utz (2004) found experimentally that priming "I" did not make individuals egoists, but induced *prosocials* to cooperate more than when "I" was not primed. She explained that this phenomenon occurred because priming "I" activated one's central social value—cooperative motivation in the case of *prosocials*. In other words, activating the *self* may make prominent an individual's (socially oriented or self-oriented) central value, and may motivate him or her to behave in certain ways that do not conflict with the central value. This means that activating the self may make *prosocials* more cooperative without developing the group identity.

Converting G_{EX} to N_{EX} may have similar effects on individuals' behaviors to those of "I" priming. Based on the aforementioned discussions, two competing scenarios can be drawn. First, if allocating private sites to individuals (N_{EX}) activates their selfish motivations, this will weaken *proselfs*' perceived group

TABLE 3.1. Summary of hypotheses.

Hypothesis 1 (H1)	Consumers will make substantially more contributions when they communicate with one another in N_{EX} than in G_{EX}.
Hypothesis 2 (H2)	The level of cooperation among individuals will be more stable when they communicate with one another in N_{EX} than in G_{EX}.
Hypothesis 3 (H3)	Consumers' level of cooperation in information exchange is likely to change minimally in N_{EX}, but to be lowered substantially in G_{EX}, as the group size increases.
Hypothesis 4a (H4a)	N_{EX} will motivate *proselfs* to cooperate more than will G_{EX}, while the level of cooperation of *prosocials* will not differ substantially between the two structures.
Hypothesis 4b (H4b)	N_{EX} will motivate *prosocials* to cooperate more than will G_{EX}, while the level of cooperation of *proselfs* will not differ substantially between the two structures.
Hypothesis 5a (H5a)	*Proselfs*' perceived group identity will be made weaker in N_{EX} than in G_{EX}.
Hypothesis 5b (H5b)	*Prosocials*' perceived group identity will be made weaker in N_{EX} than in G_{EX}.

identity more than that of *prosocials* because the orientation to selfish outcomes is the central value for *proselfs*. Eventually, this increase in selfish motivation may make cooperation in N_{EX} fragile. Second, if the structural change might activate individuals' cooperative motivation, *prosocials* will respond sensitively to the change, and their group identity will be weakened. This weakening of *prosocials*' group identity (activating the central value for *prosocials*), however, would not harm group cooperation. H5a draws from the first scenario, while hypothesis H5b draws from the second scenario.

H5a. *Proselfs' perceived group identity will be weaker in* N_{EX} *than in* G_{EX}.

H5b. *Prosocials' perceived group identity will be lower in* N_{EX} *than in* G_{EX}.

Chapter 4

Methodology

The hypotheses proposed were tested in a longitudinal experimental setting—2 (social value orientation: *prosocial* vs. *proself*) × 2 (communication structure: G_{EX} vs. N_{EX}) × 2 (group size: large vs. small) *between-subject factorial design* experiment. The primary objective of this experiment was to investigate how individuals' cooperative and defective behaviors (e.g., information provision) are shaped and conditioned differentially by the varying combinations of the experimental conditions. The two communication structures and group sizes were experimentally manipulated, and social value orientation was a measured factor. Thus, the actual communication behaviors of experimental subjects in online discussion groups with varying sizes were observed during 5 consecutive days.

Even though the many-to-many exchange of product-related information among consumers on the Internet is ubiquitous nowadays, the product categories that consumers are willing to discuss or about which they need information from others may usually be limited to a few. For example, low-involvement products, such as soft drinks, toothpaste, or laundry detergent, may not be interesting subjects about which consumers need detailed usage information from other consumers. On the contrary, it may be easily observed that consumers exchange information and opinions about such product categories as sophisticated electronics, including digital cameras and laptop computers, or medical products. Since consumers' eWOM motivations could differ substantially, depending on which product category is the subject of discussion, two pretest surveys were conducted to identify an appropriate product category to be employed in the main experiment.

PRETEST I

The purpose of this pretest was to find out the appropriate product categories that could be used in the main experiment. Thirty-five undergraduate students (n = 35) who were taking advertising/marketing courses in a large southwestern state university were recruited for this online survey. Participants recruited were asked to list, in an open-ended questionnaire, as many product categories as possible about which, they thought, "consumers including self are most likely to exchange information over the Internet." Only the answers judged appropriate and relevant, based on author's discretion, were considered. For example, answers like "electronics" were excluded because such broad categories would include many subcategories of products to which consumers might have different preferences and attitudes. Table 4.1 summarizes the results of this survey, in which product categories were rank ordered based on their frequencies of appearance.

TABLE 4.1. Rank-order of product categories.

Rank	Product Category	Frequency
1	Automobiles	12
2	Computers	9
2	Music	9
4	Travel	4
4	Clothes	4
4	Cosmetics/Accessories	4
4	Restaurants	4
8	Sports	3
8	Movie	3
8	Auctions	3
8	Books	3
12	Online Retailers	2
12	Cell Phones	2
12	Softwares	2
15	Cameras	1

The most frequently mentioned product category was "automobiles," while "cameras" was mentioned least. Because the primary concern of this survey was to identify the product categories that could be interesting for general consumers, not just for a specific group of consumers, product categories that might be of special interest for a particular group of consumers were excluded from further consideration. Thus, "clothes" and "cosmetics and accessories," which might be of more interest for females than for males, were dropped, even though they were ranked fourth. Hence, the top five product categories—"automobiles," "computers," "music," "travel," and "restaurants"—were finally selected for further consideration.

PRETEST II

The second pretest was conducted to find out which of the five product categories selected from the first pretest would be the most appropriate one to be used in the main experiment. In total, 155 (N = 155) undergraduate students who were enrolled in a large southwestern state university participated in this online survey. They were given an extra course credit as an incentive for participation. First, participants were asked to rate their level of involvement with each given product category on multiple seven-point scale items developed by Zaichkowsky (1985). Secondly, participants rated their behavioral intention to communicate with others through the Internet regarding each product category (WOM Intention) on three seven-point bipolar scale items anchored with unlikely/likely, improbable/probable, and impossible/possible.

Table 4.2 shows the descriptive statistical information of the product categories regarding product involvement and WOM intention. First, if a product category is very familiar to a particular group of consumers, while totally strange to others, using the category in the main experiment may seriously confound the experimental results because individuals' motivation to exchange information regarding the product category may be substantially differentiated by their personal product involvement. Therefore, product categories that had *smaller variances* in product involvement were considered first. Among the product categories, "computers" and "music" had the lowest standard deviations and the highest mean scores regarding product involvement. This means that consumers tended to be relatively highly involved with the two products, and the differences in product involvement levels were relatively small among individual consumers.

Second, if a consumer's WOM intention with regard to a product category is low, the product may not be a good choice for the main experiment, even though it has both a higher mean score and a lower variance in product involvement. As shown in

TABLE 4.2. Descriptive statistics of product categories.

	Mean	Variance	SD	Minimum	Maximum
Involvement					
Computer	6.48	0.56	0.75	3.0	7
Travel	5.84	1.48	1.22	1.6	7
Restaurant	5.33	1.31	1.15	1.0	7
Automobile	5.58	1.30	1.14	1.8	7
Music	6.38	0.61	0.78	3.0	7
WOM Intention					
Computer	4.89	1.40	1.18	2.0	7
Travel	5.57	1.63	1.28	1.0	7
Restaurant	5.12	1.99	1.41	1.0	7
Automobile	4.56	1.34	1.16	1.0	7
Music	5.85	1.06	1.03	2.5	7

Table 4.2, the mean of the respondents' WOM intention regarding "computers" was 4.89, while that of "music" was 5.85. This indicates that consumers may be more interested in talking about music-related products with other consumers than about computers, though their level of involvement with "computers" was as high as with "music." Based on the two criteria combined, "music" was finally selected as a product category to be used in the main experiment.

MAIN EXPERIMENT

This longitudinal experiment examines how individuals' cooperative behaviors in the context of information exchange on the Internet might be influenced by the properties of a group communication environment. Undergraduate students who were enrolled in a large southwestern university were recruited as subjects for this experiment. Participants were given an extra course credit as an incentive, and some of them were selectively provided monetary compensations as rewards, based on the overall performance of the groups to which they belonged. If a group performed well, individual members in the group were provided monetary gifts.

Experimental Procedures

Prospective participants who signed up on the participant list were contacted through an e-mail that contained a brief description of the experiment and a URL leading to one of the four experiment Web sites. By giving one *randomly* selected URL out of four to each person, participants were assigned to four experimental conditions. On the front page of the site, visitors could find a detailed description about the background, objective, and procedure of this experiment. Most parts of the instruction were the same across experimental conditions, except the sentences describing the structures of communication or group sizes. For example, in the N_{EX} condition, the instruction said, "Each participant is given a personal homepage interconnected with the others, and should make contributions by posting messages on his/her own page," while no such a statement was given in G_{EX} condition (see Appendix B).

Participants were instructed that they would be members of an online discussion group with other participants for 5 consecutive days, and were supposed to communicate with others *anonymously* (by using nicknames only) any type of information associated with "music"-related products (e.g., product description, usage experience, evaluation, news, price discount information, etc.). They were told that the main objective of this task was to construct *a useful online information pool* for potential music consumers through cooperative exchange of information and opinions, and individuals who successfully made an information pool would be given monetary gifts. It was emphasized that the monetary gifts would be given by judging the success or failure of each group as a whole, not by considering the amount of contributions made by each individual. This instruction was given to make a situation in which individuals might free ride on others' efforts: one might receive a monetary gift without making his or her own contribution, if others in the group cooperated.

In order to manipulate the *group size* conditions, they were randomly assigned either to four-person groups (small group size) or to eight-person groups (large group size). In this experiment, an eight-person group was considered a large size condition because previous social dilemma research has shown that the group size effects *asymptote* when a group contains more than eight persons (Liebrand, 1984). The amount of the monetary reward given to successfully performing groups varied depending on the group sizes: in the case of a four-person group, the total reward for a group was \$120 (= \$30 × 4), while in the eight-person case, it was \$240 (= \$30 × 8). Therefore, the payoff was the same (\$30) for each individual. Because participants were unaware not only that there were other groups, but also of how many groups would awarded the gifts, no intergroup competition situations were created.

To manipulate the communication structure conditions, half of the participants were assigned to online discussion groups, in which individuals should communicate through a *public* electronic discussion board (G_{EX}), while the other half was directed to groups, in which each individual was given his or her own personal information board accompanied by the five external links to other members' boards (N_{EX}). In this condition, no publicly shared communication space, such as an e-board, was given, but individuals were supposed to post messages on their own spaces interconnected with others (Figure 2.4). Finally, at the end of the instruction, the participants were informed that they should fill out a short online survey right after the experimental period.

On the last day of the experiment (fifth day), an e-mail message reminding the participants that they should fill out the online survey was sent out. In the survey, first, participants' *social value orientation* was measured by using the nine-item *Decomposed Games* measure (van Lange et al., 1997). Based on the results, participants were categorized into either *prosocials* or *proselfs*. Second, participants' *perceived group membership* (group identity), which would be used as a dependent variable in the later analyses, was measured by using the items developed by Yamagishi and Kiyonari (2000). Third, they were asked to rate their personal *involvement* and *knowledge* with regard to music-related products, which had been the subject discussed in the experiment. In addition, their prior experiences of participating in virtual groups of any kind were also measured. These items were used as *covariates* in the later analyses. Finally, some questions about basic demographic characteristics were asked. The total number of participants who successfully completed the online survey was 129 ($N = 129$).

Measurements

As mentioned, participants' social value orientation was measured by using a nine-item *Decomposed Games* measure (van Lange et al., 1997). Each decomposed game item showed three combinations of possible outcomes for oneself and for a hypothetical other, which were represented as value points (see Appendix B). Respondents were instructed to choose one of the possible outcomes presented. Participants who chose at least six out of the nine items in a manner consistent with their social value orientation were categorized, but those who did not meet with this criterion were excluded from the later analyses. For example, those who consistently chose the outcomes maximizing benefit for self and others more than six times were categorized as *prosocials*. People's social value orientation, measured by using the aforementioned items, has been found to be internally valid, stable for a long-term period, and free from the social desirability bias (de Cremer & van Vugt, 1999; van Vugt, 1997).

As a primary dependent variable, the level of cooperation during the experimental period was *operationalized* by calculating the number of daily contributions for each individual during the period as well as their average daily contributions. Each participant used a single nickname when posting information. After the experimental period, the messages posted under each nickname were collected, interpreted, and analyzed by two independent judges in order to make the results reliable. Based on the judges' discretion, only the information relevant to the product domain (music) was considered as a meaningful contribution, while irrelevant messages were not counted.

Each individual's contributions were distinguished further into two categories: (1) *original contribution* and (2) *reply*. An original message refers to an article posted by an individual independently, not as a response to others' messages. In contrast, a reply refers to a message created by an individual as a response to others' messages. With regard to the relevance of the messages to the "music" category and the distinction between original messages and replies, the two judges agreed on approximately 95% of decisions, and the remaining differences were resolved through discussions. Denoting the number of an individual *i*'s original messages created in a day *d* as O_{id}, his or her average daily original contributions during the experimental period *p* would be $1 / p\Sigma_{d=1}^{p} O_{id}$. A person's average daily replies were calculated in the same way. In the statistical analyses, each person's average daily contribution during the period as well as his or her actual amount of contributions per day were employed as dependent variables.

Distinguishing individual contributions into the different types is useful in two ways. First, it enables someone to consider the differing efforts that an individual makes when creating original messages or when replying to others: writing an original article may require more psychological/behavioral costs for an individual than making a response. This means that one's motivation to be an *initiator* might be substantially different from that to be a *follower*. Second, individuals' orientations to making original contributions or passive responses may be directly related to the performances of a group. If many people in a group are motivated to write original contributions, communication in the group would be very successful. In contrast, if everyone waits for others' contributions, few original contributions would be made, and the group's performance would become worse over time. Examining the two types of contributions separately allows us to consider the possible differences.

The other dependent variable to be examined is each actor's *perceived group identity*. Group identity perception was measured by using four 7-point bipolar scale items developed by Yamagishi and Kiyonari (2000). The measurement items covered four different criteria for judging the strength of a group identity: (a) *belongingness*, (b) *commonality*, (c) *closeness*, and (d) *liking* (see Appendix A). This construct's reliability score was .84.

To be used as possible covariates, participants' levels of involvement with and knowledge of the target product category were measured. Product involvement was measured by using Zaichowsky's (1985) *Personal Involvement Inventory* (*PII*) items (α = .93). Consumers' subjective product knowledge was measured by using the five items (α = .87) developed by Flynn and Goldsmith (1999). According to Brucks (1985), consumer product knowledge can be categorized into three kinds: *subjective knowledge*, *objective knowledge*, and *prior experience*. In this study, only consumers' subjective product knowledge (the extent to which consumers think about a product category) was considered, because consumers' behavioral motivation (WOM intention, in this context) may be related more with their subjective belief of their own knowledge level than the objective amount of knowledge they actually have (Alba & Hutchinson, 1987). In addition, subjects' familiarity with communicating with others through the Internet was also measured as a possible covariate by using a 7-point bipolar scale item, anchored with very familiar/not familiar.

CHAPTER 5

RESULTS

The empirical data collected through the experimental procedure were analyzed by conducting multivariate analysis of covariance (MANCOVA) and repeated-measures analysis of covariance. The initial sample size was 129, but 17 individuals with no record of activities during the 5-day experimental period were excluded. Thus, the total number of participants who were registered as members in discussion groups and completed the last online survey was 112 (male = 49; female = 63).

The effects of gender on the frequency of posting original messages were found marginally significant ($F_{(1,110)} = 3.34$, $p = .07$): male subjects made original contributions ($M_{male} = .47$) more than female subjects did ($M_{female} = .33$). Since this gender factor did not make any significant difference as a covariate in the MANCOVA model, however, it was dropped. As an initial step for examining the relationships among variables of interest, a bivariate correlation matrix was created. Table 5.1 shows the bivariate correlations among variables. As expected, the original contribution variable was significantly correlated with the reply variable [$r(112) = .25, p < .000$]. A requirement of either multivariate analysis of variance (MANOVA) or covariance (MANCOVA) is that the dependent variables to be used should be correlated to some extent, and the significant correlation found means that this requirement was met.

With regard to the covariates, product involvement was not significantly correlated with any of the dependent variables. Meanwhile, group identity, one of the dependent variables, was found to be significantly correlated with product knowledge [$r(112) = .31$, $p < .000$] and familiarity with online communication [$r(112) = .22, p < .000$]. This means that participants' perceived group identity might be influenced by their level of product knowledge as well as online communication familiarity. However, none of the covariates had significant relationships with participants' original contributions and replies. Since the information that could be obtained from these simple bivariate correlations should be quite limited, the possible effects of the covariates were also tested in the subsequent MANCOVA model.

TABLE 5.1. Correlation matrix.

	Familiarity	Original	Reply	Knowledge	Identity
Familiarity					
Original	.07				
Reply	– .01	.25*			
Knowledge	.28*	– .02	.03		
Group Identity	.22*	.03	.14	.31*	
Involvement	.10	– .01	.05	.60*	.07

*$p \leq .01$.

In this MANCOVA model, two dependent variables—one's average daily original contributions and replies—were used, while communication structure, group size, and the dichotomous social value orientation group variable (SVG hereafter) were independent factors. Among the 112 subjects, 20 individuals who were not classified either as *prosocials* or *proselfs* were not included in the statistical analyses. Therefore, the number of cases used in the model was 92 (*prosocials* = 32; *proselfs* = 60). No systematic pattern between SVG and subjects' gender distribution was found ($\chi^2 = 2.71, p = ns$). As in the correlation results, product involvement and knowledge did not have significant effects on dependent variables, thus they were dropped from the model. Unlike the correlation results, however, it was found that participants' degree of familiarity with online communication had statistically significant effects on their frequency of making original contributions. Therefore, the variable of online communication familiarity was included as a covariate in the model.

Table 5.2 summarizes the multivariate and univariate significance test results. The second column of the table shows

TABLE 5.2. Multivariate and univariate significance test results.

	Multivariate	Original	Reply
Familiarity	2.88***	5.50*	.00
Structure (S)	6.00**	12.06**	.29
Group Size (G)	1.83	3.70***	.26
Social Value Group (V)	.06	.11	.01
S × G	2.55***	5.12*	.10
S × V	2.89***	3.84***	.84
G × V	.31	.06	.46
S × G × V	2.62***	5.30*	.38

MANCOVA *df* = 2/82. Univariate *df* = 1/83.
$^*p \leq .05$. $^{**}p \leq .01$. $^{***}p \leq .10$.

the multivariate *F* ratios for the independent factors and the interactions between them. The effects of the covariate—familiarity—on the linear composite dependent variate were marginally significant ($F_{(2,82)} = 2.88$, $p = .06$), which means that the number of original contributions as well as replies might be influenced by participants' degree of familiarity in communicating with others over the Internet. That is, the result indicates that consumers who are familiar with using the online communication channels (e.g., chat room, e-board, instant messenger) may tend to be more cooperative than those who are not.

After controlling the effects of individual variations in familiarity with online communication, the effects of communication structure (N_{EX} vs. G_{EX}) on individual communication behaviors could be revealed more clearly. As shown in Table 5.2, it was found that participants' communication behaviors, including writing original messages and replies, were significantly differentiated by the two structures ($F_{(2,82)} = 6.00$, $p < .000$). On the other hand, the main effects of group size ($F_{(2,82)} = 1.83$, $p = ns$) and SVG ($F_{(2,82)} = .06$, $p = ns$) were not statistically significant. These findings imply that individuals' communication behaviors may be influenced by the communication structures but not directly by group sizes or personality characteristics.

In contrast to the main effects, group size and SVG had meaningful interaction effects, together with the structure variable. The effects of the structure by group size interaction term on the dependent variate were marginally significant ($F_{(2,82)} = 2.55$, $p = .08$), and the effects of the structure by SVG were also marginally significant ($F_{(2,82)} = 2.89$, $p = .06$). These marginally significant interaction effects seem to result largely from the insufficient statistical power due to the relatively small sample size employed, but also provide rationales for examining further the interaction effects in the subsequent univariate analyses.

The group size by SVG group interaction effects were not found, but the three-way interaction effects between the independent factors were marginally significant ($F_{(2,82)} = 2.62$, $p = .08$), which requires further analysis.

The third and fourth columns show the univariate F ratios for two dependent variables respectively. Table 5.2 clearly shows that it was only the original contribution variable that was influenced by the independent factors, while the reply variable was not influenced by any factor. First of all, it was found that the average daily frequency of making original contributions was influenced by participants' level of online communication familiarity ($b = .04$; $F_{(1,83)} = 5.50$, $p < .05$). This means that individuals who had many experiences of communicating with others over the Internet tended to write a higher number of original messages than did those who had not. In contrast, this was not the case in terms of the average daily frequency of making replies ($F_{(1,83)} = .00$, $p = ns$).

Having controlled the familiarity effects, it was found that the frequency of participants' original contributions was substantially

Table 5.3. Adjusted means and standard errors.

	N_{EX}			G_{EX}		
	Large	Small	Total	Large	Small	Total
Original						
Prosocials	.91 (.11)	.38 (.10)	.64 (.07)	.08 (.10)	.32 (.23)	.20 (.12)
Proselfs	.54 (.08)	.37 (.08)	.46 (.06)	.43 (.08)	.24 (.08)	.33 (.06)
Total	.72 (.07)	.38 (.06)	.55 (.05)	.25 (.07)	.29 (.12)	.26 (.07)
Reply						
Prosocials	.73 (.22)	.38 (.20)	.56 (.14)	.64 (.20)	.60 (.45)	.62 (.25)
Proselfs	.68 (.17)	.76 (.17)	.72 (.12)	.49 (.17)	.47 (.17)	.48 (.12)
Total	.71 (.14)	.57 (.13)	.64 (.09)	.57 (.13)	.53 (.24)	.55 (.14)

Note. Values enclosed within parentheses represent standard errors.

differentiated by the two exchange structures. The adjusted mean score of the N_{EX} condition (M_{NEX} = .55, *SE* = .05) was approximately two times higher than that of the G_{EX} condition (M_{GEX} = .26, *SE* = .07; $F_{(1/83)} = 12.06$, $p < .000$). Hypothesis 1 predicted that individuals would make more contributions in N_{EX} than G_{EX}. Given the results, therefore, H1 was confirmed. The main effects of group size on the level of original contribution were marginally significant ($F_{(1,83)} = 3.70$, $p = .06$), but the direction of the effects was opposite to what previous studies predicted: participants made more original contributions when their group was large (M_{large} = .49, *SE* = .05) than small (M_{small} = .33, *SE* = .07). This counterintuitive finding will be discussed further later.

In this model, the main effects of SVG were not found to be statistically significant both on original contributions ($F_{(1,83)} = .11$, $p = ns$) and on replies ($F_{(1/83)} = .01$, $p = ns$). This implies that individuals' cooperative behaviors may be influenced more by the social environmental settings given than by their personality characteristics. H2 predicted that cooperative information exchange would be more *longitudinally stable* in N_{EX} than in G_{EX}. In other words, it was hypothesized that the level of contributions in G_{EX} would decrease during the period, while those in N_{EX} would remain relatively stable. To test this hypothesis, two repeated-measures analyses of covariance were conducted by using original contribution and reply, respectively, as dependent variables. As in the previous MANCOVA model, participants' degree of familiarity with online social communication was incorporated as a covariate in these analyses. Because SVG was not examined in these models, the total number of cases examined was 112.

An important assumption of repeated-measures analysis is that the variances of within-factors as well as across between

TABLE 5.4. Repeated measures ANCOVA results.

	SS	*df*	*MSE*	*F*
Original				
Period	4.73	3.03	1.56	2.44*
Period × Familiarity	1.21	3.03	0.40	0.63
Period × Structure	4.88	3.03	1.61	2.52*
Period × Group Size	1.84	3.03	0.61	0.95
P × S × G	1.84	3.03	0.61	0.95
Error	207.18	324.29	0.64	
Reply				
Period	12.66	3.50	3.62	3.18**
Period × Familiarity	2.14	3.50	0.61	0.54
Period × Structure	3.06	3.50	0.87	0.77
Period × Group Size	3.38	3.50	0.96	0.85
P × S × G	6.49	3.50	1.86	1.63
Error	426.43	374.61	1.14	

Note. df = Greenhouse–Geisser adjusted degrees of freedom.
*$p \leq .10$. **$p \leq .05$.

conditions should be roughly equal, which is called "sphericity." If this assumption were violated, the test's statistical power would decrease substantially, which leads to the increase of Type II error. Mauchly's test statistic W enables someone to judge whether the sphericity assumption is violated, and with the current data, it was found that the sphericity assumption was violated severely for both original contribution (W = .42; χ^2 = 90.82, $p < .01$) and reply (W = .73; χ^2 = 33.81, $p < .01$). Due to the significant violation of sphericity, this study employed Greenhouse–Geisser's adjusted degrees of freedom (*df*), which is the most conservative one, to test the effects. Table 5.4 summarizes the results.

With regard to the original contribution, the results show that the main effects of "period," the within-factor ($F_{(3.03, 324.29)} = 2.44$, $p = .06$), and the period-by-structure interaction effects ($F_{(3.03,}$

$_{324.29)} = 2.52$, $p = .06$) were marginally significant. This means that participants' level of making original contributions fluctuated significantly during the experimental period, and the fluctuation patterns differed substantially between the two structural conditions. Meanwhile, in the case of "reply," the main effects of "period" were only statistically significant ($F_{(3.50,\ 374.61)} = 3.18$,

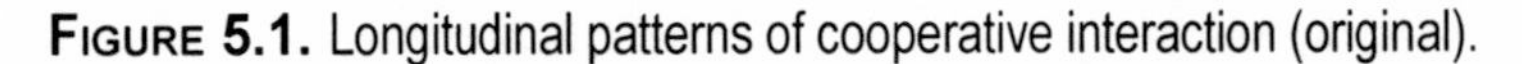

FIGURE 5.1. Longitudinal patterns of cooperative interaction (original).

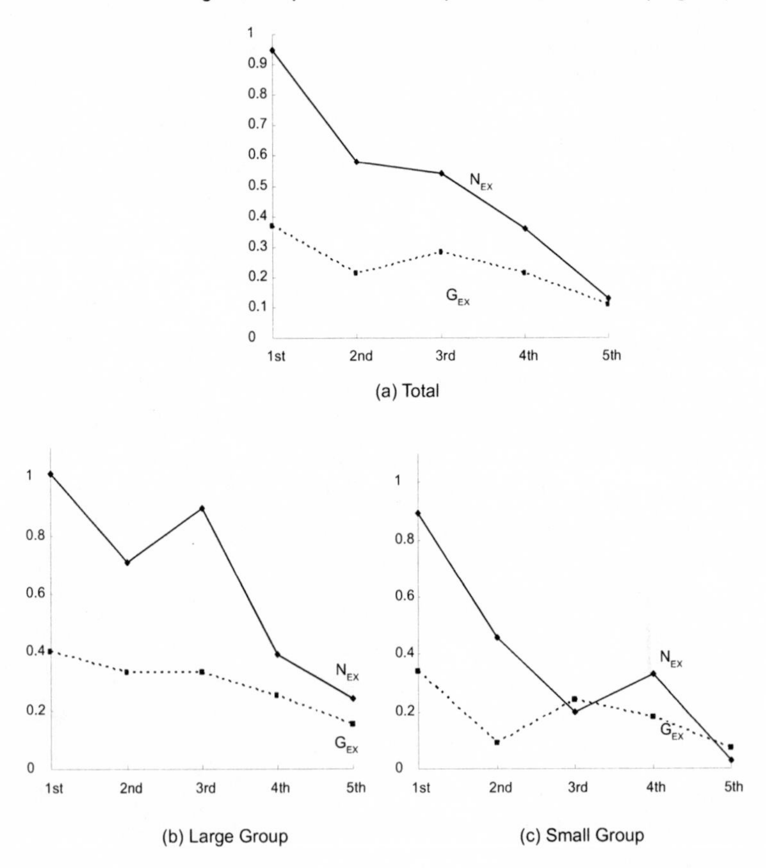

$p < .05$), while the other between-subject factors did not make any difference.

In order to see further how the fluctuating patterns of original contributions were differentiated by the structures, the contribution frequencies during the 5 days were examined across the two structural conditions. Figure 5.1 graphically illustrates how the longitudinal patterns of participants' original contributions were differentiated by the structures. As shown in Figure 5.1(a), the level of original contributions was initially very high, but dropped rapidly in N_{EX}, while the contribution level remained relatively stable in G_{EX}. Because this pattern is opposite to the predicted one, H2 was not supported by the results. This failure to confirm H2 seems partly due to the short experimental period. Since participants knew that the experiment would end after 5 days, their motivation to make contributions might have faded as the experiment moved toward the end.

For this reason, the initial high level of contributions could not be maintained, but converged to that of G_{EX}. Figure 5.1 (b) and (c) shows the interaction patterns across group size conditions. Although both graphs show a decreasing pattern, the rate of decrease in cooperation seems relatively smaller when the size of a group is large rather than small. This implies that cooperative interaction in N_{EX} may be stabilized if a large number of actors in the network interact for a longer period of time. This possibility will be discussed further in the discussion chapter.

H3 was about the interaction between communication structure and group size. It predicted that participants' contributions in N_{EX} would increase or change minimally as the group size increased, while those in G_{EX} would decrease as the group got bigger. Table 5.3 shows that people's contribution level in G_{EX} was not differentiated much by the group size factor. The difference in the number of original contributions was minimal between large and small conditions (M_{large} = .25, *SE* = .07; M_{small} = .28, *SE* = .12). This small difference between large and small group conditions seems to be because the contribution level in G_{EX} was so low that the additional decrease caused by the group size factor became unnoticeable. In contrast, participants in N_{EX} made a larger number of original contributions in a large than small group situation (M_{large} = .72, *SE* = .07; M_{small} = .38, *SE* = .06; $F_{(1,83)} = 5.12$, $p < .05$). Figure 5.2 (a) shows the interaction pattern graphically. Because it was found that cooperation in N_{EX} was positively affected by the group size factor, though the negative effects of group size on G_{EX} were minimal, H3 was partially confirmed.

H4a (responsibility-effects hypothesis) suggested that N_{EX} would motivate *proselfs* to cooperate, more than it would motivate *prosocials*, by increasing the level of personal responsibility imposed on individuals. On the contrary, H4b (efficacy-effects hypothesis) predicted that N_{EX} would induce *prosocials* to respond more sensitively than *proselfs* by enhancing self-efficacy

FIGURE 5.2. Patterns of interaction (structure × group size and structure × SVG).

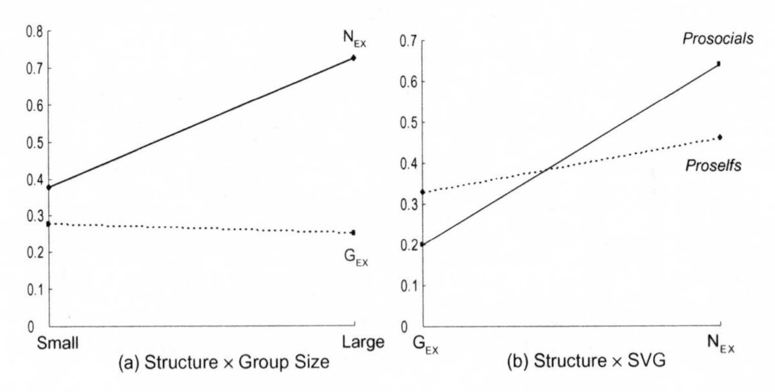

of contribution. The results of MANCOVA analyses (Tables 5.2 and 5.3) revealed that *prosocials* responded more sensitively to the structural change than did *proselfs*, which confirms H3b—the efficacy-effects hypothesis ($F_{(1,83)} = 3.84, p = .053$).

Participants who were classified as *prosocials* contributed more in N_{EX} ($M_{nex} = .64, SE = .07$) than in G_{EX} ($M_{gex} = .20, SE = .12$), while the contribution level of those classified as *proselfs* was not differentiated significantly by the communication structures ($M_{nex} = .46, SE = .06$; $M_{gex} = .33, SE = .06$). Figure 5.2 (b) illustrates graphically the interaction pattern between the two communication structures and SVG. The results imply that the increase in cooperation in N_{EX} may not be because individuals become sensitive to personal responsibility and social pressures, but because they can enhance the visibility of their contributions in N_{EX}. Based on the results, therefore, H4b was confirmed, while H4a was not.

Although not hypothesized, the three-way interaction effects on original contributions were found to be statistically significant ($F_{(1, 83)} = 5.30, p < .05$). As shown in Figure 5.3 (a), the structural change from G_{EX} to N_{EX} made little difference in how many original contributions the *prosocials*' made, but made more difference in *proselfs*' behaviors, when the group size was small. Figure 5.3 (b), however, shows that the structural change caused more difference in *prosocials*' behaviors than in *proselfs*' behaviors, when the group was large. In a small-group situation, changing G_{EX} to N_{EX} might increase the amount of personal responsibility imposed on each individual, while making no meaningful difference in contribution efficacy. Thus, *proselfs*, who tended to be more sensitive to social pressure, might become more cooperative. On the contrary, in a large-group situation, changing G_{EX} to N_{EX} would increase the visibility of one's contribution, which might become an incentive for *prosocials* to make more contributions. This implies that the effects of group size could be positive, particularly for *prosocials* in N_{EX}.

Finally, H5a and H5b were about the impact of the structural change from G_{EX} to N_{EX} on the formation of group identity. H5a stated that N_{EX} would make individual actors more oriented to their selfish interests than to group goals, and this impact would be revealed as the weakened group identity of *proselfs*. On the contrary, H5b expected that N_{EX} would not make individuals egoists, but would activate their central social values. In this case, it was predicted that *prosocials*' group identity would be weakened by the structural change, which would not decrease their cooperative motivations. To test the hypotheses, an ANCOVA model was constructed. This model included participants' perceived group identity as a dependent variable, with communication structure and SVG as independent variables. In addition, respondents' familiarity with online social communication and product knowledge level were included as covariates because they were highly correlated with perceived group identity as shown in Table 5.1.

FIGURE 5.3. Three-way interaction patterns (structure × group size × SVG).

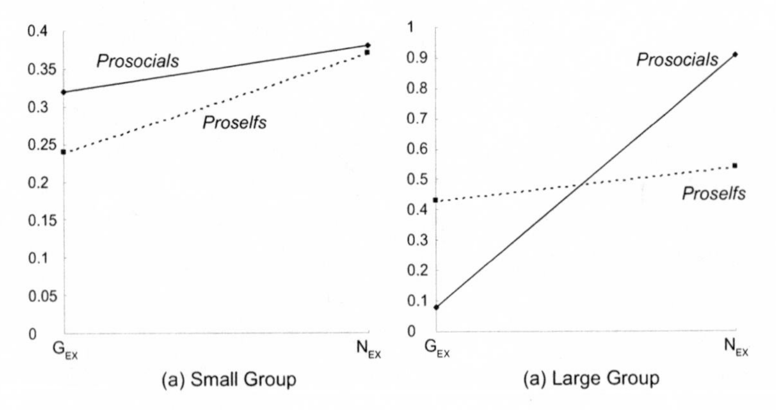

Table 5.5. ANCOVA results.

	SS	*df*	*MSE*	*F*
Familiarity	5.11	1	5.11	3.10*
Product Knowledge	5.45	1	5.45	3.31*
Structure (S)	2.03	1	2.03	1.23
SVG	0.14	1	0.14	0.09
S × SVG	3.74	1	3.74	2.27
Error	141.74	86	1.65	

*$p \leq .10$.

Table 5.5 shows the significance test results. First, it was found that participants' perceived group identity was marginally associated with their degree of familiarity ($b = .11$; $F_{(1,86)} = 3.10$, $p = .08$) as well as product knowledge ($b = .18$; $F_{(1,86)} = 3.31$, $p = .07$). This means that consumers may have a relatively strong perception of membership of a product-related discussion group when they are experienced with such online communications and knowledgeable about the product category. After controlling for the effects of covariates, unfortunately, none of the independent factors was found to significantly affect the dependent variable. Although the hypothesized interaction effects were not statistically confirmed, Figure 5.3 provides some useful clues for inferring the effects of the structural change on the formation of group identity.

As shown previously, *prosocials*' group identity became weaker in N_{EX} ($M_{NEX} = 3.04$, $SE = .30$) than in G_{EX} ($M_{GEX} = 3.80$, $SE = .37$), while *proselfs*' group identity was not changed much by the structural shift ($M_{NEX} = 3.57$, $SE = .24$; $M_{GEX} = 3.45$, $SE = .24$). The fact that *prosocials*' cooperative behaviors were higher in N_{EX} than in G_{EX} implies that the changing communication structure from G_{EX} to N_{EX} might not activate individuals' egoistic motivations, but may instead make cooperation

FIGURE 5.4. Change of group identity.

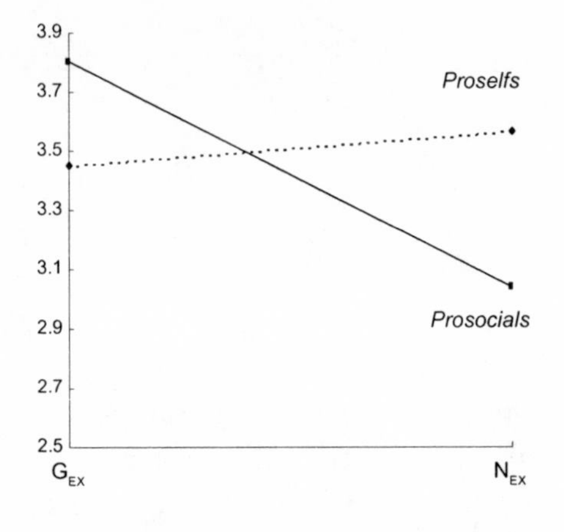

among individuals less dependent on their shared group identity. Because it is particularly difficult to build a solid group identity in an online environment where many anonymous individuals interact, this result indicates that N_{EX} can be a viable solution to foster cooperation without constructing group membership.

CHAPTER 6

DISCUSSION

The empirical findings of this study demonstrate that consumers' online communication behavior is partly a function of the *structural* properties of the communication environment. Previous studies dealing with virtual communities have understood a consumer's participation in online groups or communities as a product of his or her personal decision, a decision that is largely unrelated with those of others. For example, Bagozzi and Dholakia (2002) proposed that one's participation in a virtual community is an outcome of the "we-intention" that the person has, which is influenced by various cognitive and emotional antecedents. Also, Hennig-Thurau, Gwinner, Walsh, and Gremler (2004) conceptualized an individual's eWOM intention as a function of multiple personal utilities. Although these approaches illuminated some important psychological aspects of the social behavior, they were unable to reveal how individuals behave adaptively when responding to others' behaviors in the multiagent interaction system.

H1 proposed that N_{EX} would motivate individual actors to make substantially more contributions than G_{EX} would. As reported in Table 5.1, it was found that subjects in N_{EX} provided significantly more original information than those in G_{EX}. This finding implies that an individual's participation in a virtual community is not an outcome of socially isolated decision making, but a social behavior shaped and conditioned by the communication environment consisting of other actors. That is, a person's participation decision in a multiagent communication setting can be understood as an *adaptive* response to his or her immediate social surroundings in which N-1 others also do the same, as Schelling (1978) pointed out earlier.

H2 predicted that communication in N_{EX} would be longitudinally more stable than that in G_{EX}, which means that individual actors' noncooperation would increase dramatically in G_{EX} as time went by, while this would not be the case for N_{EX}. Unfortunately, this hypothesis was not supported by the experimental results. As mentioned before, however, this failure to confirm the hypothesis seems to emanate from the limitations of the experimental setting. First, the experimental period—5 days—was too short to observe the difference in the actual longitudinal stability between N_{EX} and G_{EX}. Because participants were aware that the experiment would come to an end after 5 days, they might not have any reason or incentive to continue to contribute to the online discussion group until the last minute. As a result, the initial level of contribution in N_{EX} could not be sustained, and it dropped rapidly toward the end of the period. In contrast, communication in G_{EX} remained relatively stable, because its initial contribution level was too low. This means that it is necessary to observe communication in both N_{EX} and G_{EX} for a longer period of time to get a reliable answer for this question.

Secondly, Figure 5.1 (b) and (c) reveal another possibility. Although the three-way interaction effects (structure x period x group size) were not statistically confirmed, the two graphs show a meaningful difference: the rate of decrease in Figure (b) appears to be relatively slower than that in Figure (c). This indicates that cooperation in N_{EX} may be more sustainable when the group size is large than when it is small. Oliver and Marwell (1988) argued that under some circumstances, group size may affect positively collective action because the probability of having cooperators in a group and of reaching a critical mass would be higher in a large group than in a small one. Applying this notion to the current context of sustaining N_{EX}-based communication, there may be a higher probability of success in a large-sized group than in a small one. Yamagishi and Cook (1993) also pointed out that small-group cooperation can be more vulnerable to the free-rider problem because it only takes a few noncooperators to cause serious trouble in maintaining the cooperative process. Conversely, cooperation in a large-sized group can be sustained even if many actors free ride because it is relatively more probable that there are cooperators necessary for maintaining the group.

Then, under what circumstances does the group size factor positively affect the cooperative process? What factors determine whether the group size factor will have positive or negative effects on cooperation? This question is directly related to H3. In H3, it was stated that the level of contributions in N_{EX} would increase or change minimally as the group gets larger, while in G_{EX} they would decrease accordingly. Previous discussions regarding the negative effects of group size were based on the notion that an individual's responsibility for cooperation and self-efficacy of contribution would decrease altogether in a large-sized group (Kerr, 1989; Olson, 1965). This inference was based on the conception of a group as an aggregate of individuals, which has no particular structure of social exchange.

The rationale underlying H3, however, was that there can be certain structures facilitating or hindering the multiagent interactions, and N_{EX} was proposed as an alternative structure maintaining both personal responsibility and contribution efficacy for each individual, regardless of the size of the group involved. The experimental results supported this prediction: in N_{EX}, subjects' average daily original contribution level increased as the group size increased, while that was not the case for G_{EX}. This means that the direction of effects of group size may change depending on the types of communication structure. In turn, this demonstrates that networking individuals is superior to simply grouping them for fostering cooperative information exchange among a large number of individuals. Furthermore, this finding illuminates that N_{EX} may be longitudinally stable when the size of a group is large, though this claim was not empirically confirmed in this study. Further research is needed to test this claim.

H4a and H4b were two competing statements about the underlying psychological effects of changing G_{EX} to N_{EX}. H4a suggested that N_{EX} would make *proselfs* more cooperative by making them sensitive to the increased personal responsibility. On the contrary, H4b predicted that N_{EX} would motivate *prosocials* to be more cooperative by increasing the visibility of their contributions. The experimental results supported H4b instead of H4a. This means that N_{EX} increased individuals' cooperative motivation primarily by creating a more efficacious communication environment, not by imposing more responsibility on them. In other words, changing G_{EX} to N_{EX} would attract individuals who are socially motivated rather than those with individual orientation.

Individuals often cooperate with others when their own selfish interests meet with collective interest. The cooperation among such egoists should, however, be unstable, because any subtle change in the incentive structure (e.g., decrease in rewards or increase in losses) could activate individuals' selfish motivations, which eventually leads to the end of cooperation. Conversely, cooperation among socially oriented individuals, *prosocials*, may be relatively insensitive to the environmental changes, because *prosocials* are inherently motivated to cooperate with others. This implies that motivating *prosocials* may be a more reliable and safer way to construct and maintain a virtual community than is the opposite case. In this sense, the results that *prosocials* responded more sensitively to the structural change from G_{EX} to N_{EX} imply that N_{EX} can be a reliable solution to the free-rider problem in virtual communities.

In a similar vein, finally, H5a and H5b made two opposite predictions about the impact of the structural change from G_{EX} to N_{EX} on individuals' perceived group identity. H5a stated that changing G_{EX} to N_{EX} would weaken *proselfs*' group identity, while H5b suggested that the structural change would weaken *prosocials*' group identity. As mentioned before, weakening proselfs' group membership may be hazardous to the group, because *proselfs* may be easily tempted to free ride if the group's solidarity becomes weak. On the contrary, weakening *prosocials*' group membership may not cause a serious problem for the group, because their cooperative motivation is not heavily dependent on the group's solidarity.

Although not statistically confirmed, the results show that N_{EX} weakened *prosocials*' perceived group identity more than that of *proselfs*. Given that *prosocials* contributed more in N_{EX} than G_{EX}, the decrease of *prosocials*' perceived group identity in N_{EX} indicates that N_{EX} could possibly promote cooperation among individuals without developing a shared group identity. The implication of this notion is important for virtual community developers because making individuals perceive themselves as members of a group is very difficult, particularly when a large number of anonymous individuals participate in the community. Thus, in such a situation, changing communication structures would be a more viable means for facilitating cooperative interactions among participants.

Chapter 7

Implications and Conclusion

Theoretical Implications

Since the first appearance of *the two-step flow of communication* model (Katz & Lazarsfeld, 1955), information diffusion has been believed to be a social process in which individuals called *opinion leaders* play a crucial role as "intervening factors between the stimuli of the media and resultant opinions, decisions and actions" (p. 32). Inspired by the sociological portrayal of the mass communication process, scholars in the field of marketing and advertising have long been interested in the proactive role of consumers as information transmitters. However, in most previous studies dealing with consumer WOM, information flow has been viewed as a *unidirectional* process through which opinion leaders transmit information to the rest of the information consumers (Gatignon & Robertson, 1998).

This perspective is misleading, for three reasons. First, interpersonal word of mouth can occur *horizontally* (e.g., information sharing, reciprocal information exchange) as well as *vertically* (Robinson, 1976). Second, a person's role either as an information transmitter or receiver may change, depending on situations (Lin, 1971). Finally, portraying the information receiver as an end user of information shadows the fact that an individual who has received information may have two alternative courses of action: (1) *Personal utilization* of information without disclosing it, and (2) *transmission* of information to others. Information is not a disposable item to be searched, acquired, and consumed by an individual in isolation. Instead, it is an item to be transferred and shared among multiple individuals. If most actors in a society are merely information consumers, as mainstream economics and the literature of consumer information search behavior portrayed them (for review, see Peterson & Merino, 2003), then diffusion of information will not occur (Valente, 1995). The occurrence of information diffusion requires that one should be a transmitter of information at a point.

Even though we accept the premise that there should be individual variations in the extent of transferring information to others, the conventional attempts to polarize information givers and receivers have made researchers miss the fact that an individual is an essential part in the nexus of communication. This means that consumers who initiate WOM communication may not necessarily be opinion leaders (Richins, 1983). As shown in the findings of the current study, individuals with socially oriented values are more likely to provide information for others than are those with an individual orientation. This, however, does not mean that the socially oriented people—*prosocials*, so to speak—are necessarily opinion leaders. In other words, opinion leadership is not a necessary condition for consumer WOM to occur.

Recognizing that information transmission is not a trait of an individual, but a context-dependent *behavior*, on the other hand, some diffusion researchers have investigated primarily the social structural factors affecting individuals' WOM behaviors. With regard to the social contagion process, two competing hypotheses—*cohesion* vs. *structural equivalence*—have long been under debate in this approach. The cohesion hypothesis proposes that social contagion (e.g., innovation diffusion, the spread of disease) occurs primarily through interactions and communications among individuals proximate in a social structure (e.g., close friends). Meanwhile, the structural equivalence hypothesis suggests that social contagion occurs through social comparison processes among individuals who are equivalent in terms of positions in a social structure. Burt (1987) tested these two hypotheses in the context of physicians' adoption of tetracycline, and found that the structural equivalence hypothesis better explained the social contagion process than did cohesion.

Although those diffusion-related studies have illuminated the effects of social structures on the contagion process, they have not illuminated how the personal attributes and social structural characteristics of the communication environment systematically interact with one another. By portraying diffusion as a process without conflict, researchers have not seriously asked under what social structural circumstances an individual's motivation to communicate is in conflict or coordination with those motivations of others (Bonacich, 1990). As *social exchange* theorists have argued, a behavior like transferring resources to others is made on the intersection of *individual motivation*, which entails consideration of the relevant benefits or losses, and the *social structural constraints* imposed (Emerson, 1976; Foa & Foa, 1980). Without either of the two, consumer WOM cannot occur (Frenzen & Nakamoto, 1993).

The Internet has created an incredibly efficient environment for multiagent information exchange. In this N-person interaction situation, one's WOM decision is an act neither originating solely from personal motivations or interests, nor determined by the ongoing social relationships with others, as assumed in many network analyses. Instead, a person's WOM decision is an *adaptive* response made on the intersections between his or her personal predispositions and what N-1 others do. In short, what should be focused on to grasp consumer WOM in the multiagent interaction setting are the combined effects of personal attributes and situational dynamics on individuals' motivation to transfer information and knowledge to others. This study is an attempt to understand consumer's WOM decisions on the Internet as an adaptive dynamic between individual characteristics and the social-structural environment.

LIMITATIONS AND ISSUES FOR FUTURE RESEARCH

The current study attempted to reveal that consumer's electronic word of mouth (eWOM) is a product of varying combinations of a personal characteristic—social value orientation—and the properties of a group communication environment, like exchange structure and group size. A major finding of this study is that "networking" is superior to "grouping" for motivating individuals to engage in communication with others. However, the network considered here is a special one, a complete network, in which every node is connected with all the others (density = 1.00). Obviously, a network's density of connections may vary from zero (no connection) to one (complete connection), and a network's density level will decrease rapidly as the number of nodes in the network increases[1] (Wasserman & Faust, 1994). This means that it will be extremely difficult to make a completely connected network when the size of a group is very large. Therefore, more empirical studies should be conducted to see how individuals' communicative motivation is influenced by the varying proportions of connections.

In addition to the varying degrees of network density, other properties of a network can be interesting subjects to research. For example, the distribution of connections in a network may vary from a completely *random* distribution to a completely centralized distribution (Freeman, 1979; Skvoretz, 1985). In a perfectly random network, every individual is connected to the same number of others. In a perfectly centralized network, conversely, one person is connected to all the others, while the N-1 others are not connected with one another. The distributions of connections in most real-world networks will be in between the two extreme cases, and Barabasi and Albert (1999), for example, found that the Internet is a highly centralized network (scale-free network, in their terms), in which a small number of nodes have most connections. How will consumers' eWOM decisions be influenced by the distribution patterns of connections? Answering this question would be valuable for more systematically understanding individuals' communication behaviors in the emerging interpersonal networks of mediated communication.

This study investigated the individual-structure interaction in a 5-day longitudinal experiment, which was quite limited in observing the actual *evolutionary* process of cooperative communication in virtual communities. A longer experimental investigation will be necessary for further research of the life cycles of virtual groups or communities. Furthermore, since the main concern was the quantitative information of individual contributions, the actual contents of messages communicated were not a focus of this study. Undoubtedly, consumers' communicative motivation may change, depending on the types of issues discussed (e.g., political issues, cultural issues). Examining qualitatively the contents of what consumers communicate in a virtual community setting may produce richer knowledge of the online social communication process. Finally, researchers should investigate the influence of personal characteristics (other than social value orientation) on consumer's eWOM decisions and behaviors.

MANAGERIAL IMPLICATIONS

Virtual social interaction among consumers is already an indispensable element in explaining the nature of the interactive media environment. The communication goals of companies in this environment cannot be achieved without understanding and adapting to the dynamics of consumer-to-consumer interactions (Hagel & Armstrong, 1997). Adequate knowledge about consumer eWOM behaviors will allow companies to use the virtual consumer communities as a remarkably cost-effective channel for disseminating information and interacting with target consumers. Without this knowledge, the virtual social collectivities will turn to uncontrollable noises or obstacles for companies' communication efforts (Rosen, 2000).

As mentioned before, a virtual consumer community can be useful for companies in two different ways. First, companies may reach their target consumers in relatively unobtrusive ways (company-to-consumer connection). For a company trying to introduce a new digital camera product, for example, virtual consumer communities may serve as an efficient channel through which information about the product could be disseminated with little cost. Secondly, consumers can express their opinions and ideas regarding products and services in a virtual community (consumer-to-company connection). Collecting those "unbiased" consumer opinions about products and services should be very useful for a company's managerial decision-making process.

Recognizing the potential of virtual consumer communities, a variety of commercial and noncommercial organizations have already tried not only to create channels like electronic bulletin boards, but also to organize virtual consumer communities or forums to promote consumers' eWOM activities. The problem, however, is that there is no reliable knowledge or guideline about how to organize and manage such virtual consumer communities successfully. Most studies dealing with consumer behavior on the Internet have concentrated on the psychological stimulus-and-response process between humans and computers and/or messages as a proxy of companies. As a result, little academic research has been conducted to envision the elements needed for growing the virtual consumer communities.

The findings of this study reveal only a small part of the knowledge that is necessary for building and prospering a virtual community. For example, imagine that a company planning to organize an online consumer group to learn consumer opinions about a new product before introducing it to a market. What should the company do to motivate consumers in the group to contribute actively? In other words, how can the participants' free-riding motivation be reduced? According to this study's results, allocating a private homepage (e.g., blog) to each consumer will be a better solution for promoting product-related conversations than simply creating a public e-board. Over the past several decades, social dilemma researchers have identified many other solutions to promote individuals' cooperative behaviors (Kollock, 1998; Messick & Brewer, 1983). Applying the knowledge established to the context of virtual social interactions would produce a more concrete body of knowledge that would provide rules and guidelines for appropriate managerial decision making, though there are numerous ethical and practical problems ahead, waiting to be solved.

CONCLUSION

During the past decade, the Internet has received tremendous scholarly attention in the field of advertising research. Numerous research articles and book chapters dealing with the applications of the Internet for communicating with consumers have been published. In spite of the numerous intellectual efforts, unfortunately, the Internet still remains an elusive medium. Even after decades of research, more questions than answers remain. This ironical situation seems to result partly from the myopic perspective that has been dominant in this area of research, which views the Internet as an "interactive" medium facilitating only two-way communication between an individual consumer and a company.

There is no doubt that the Internet is an *interactive* medium. The point is that *interactivity* should be a broader concept, encompassing not only the dyadic interaction between two entities but also the interaction among multiple entities. Clearly, the uniqueness of the Internet comes from its ability to enable individuals to interact socially with multiple others beyond existing social and physical boundaries. Previous interactivity studies have not explicitly considered this social motivation of consumers on the Internet, and thus have failed to recognize the true potential of this new medium.

This research is an attempt to shed light on the social dimension of interactivity in the context of market information exchange among consumers on the Internet, which has been virtually neglected by advertising researchers. It becomes increasingly clear that the power of ordinary consumers in the CME will continue to grow, as the mobile communication technologies, combined with the Internet, are widely adopted. In this situation, grasping the process of consumer-to-consumer communication is no longer a matter of choice for companies. Consumers are motivated and able to create and communicate information with others, which currently does and will continue to have an impact on the market process. A key to grasping and adapting successfully to this new environment is to find ways to coexist with consumers in this emerging "networked" communication environment.

ENDNOTE

1. Network density can be expressed formally as follows: $\Delta = \Sigma^n_{i=1}\Sigma^n_{j=1}x_{ij} / \max \Sigma^n_{i=1}\Sigma^n_{j=1}x_{ij}{}^*$ where x_{ij} is the relationship value between node *i* and *j*, and $x_{ij}{}^*$ is the maximum possible relationship value. If the relationship value indicates just the existence/absence of a link as 0 or 1, the formula becomes $L/n(n-1)$, where *L* is simply the number of existing links, and *n* is the number of nodes.

Appendix A

Additional Tables and Figures

Table A-1. Multivariate results of repeated measures analyses.

	Wilks' Λ	*F*	Hypothesis *df*	Error *df*
Original				
Period	.81	6.08**	4	104
Period × Familiarity	.98	0.56	4	104
Period × Structure	.84	4.98**	4	104
Period × Group Size	.96	1.00	4	104
P × S × G	.97	0.92	4	104
Reply				
Period	.91	2.72*	4	104
Period × Familiarity	.98	0.52	4	104
Period × Structure	.97	0.89	4	104
Period × Group Size	.97	0.69	4	104
P × S × G	.94	1.75	4	104

*$p \leq .05$. **$p \leq .01$.

TABLE A-2. Means and standard errors (repeated measures analyses).

	N_{EX}			G_{EX}		
	Large	**Small**	***Total***	**Large**	**Small**	***Total***
Original						
1st	1.01 (.18)	.89 (.18)	.95 (.13)	.40 (.19)	.34 (.21)	.37 (.14)
2nd	.71 (.11)	.46 (.12)	.58 (.08)	.33 (.12)	.09 (.13)	.21 (.09)
3rd	.89 (.14)	.20 (.14)	.54 (.10)	.33 (.14)	.24 (.16)	.28 (.11)
4th	.39 (.12)	.33 (.12)	.36 (.09)	.25 (.13)	.18 (.14)	.21 (.09)
5th	.24 (.07)	.03 (.07)	.13 (.05)	.15 (.07)	.07 (.08)	.11 (.05)
Total	.64 (.06)	.38 (.06)	.51 (.05)	.29 (.07)	.19 (.07)	.24 (.05)
Reply						
1st	.50 (.16)	.53 (.16)	.51 (.11)	.40 (.16)	.33 (.18)	.36 (.12)
2nd	.77 (.26)	1.38 (.26)	1.07 (.19)	.74 (.27)	.52 (.31)	.63 (.20)
3rd	1.04 (.20)	.67 (.20)	.85 (.14)	.68 (.20)	.69 (.23)	.68 (.15)
4th	.84 (.19)	.53 (.19)	.69 (.14)	.43 (.20)	.43 (.22)	.43 (.15)
5th	.38 (.17)	.49 (.17)	.44 (.12)	.33 (.18)	.58 (.20)	.45 (.13)
Total	.71 (.12)	.72 (.12)	.72 (.08)	.52 (.12)	.51 (.13)	.51 (.09)

Note. Values enclosed in parentheses represent standard errors.

FIGURE A-1. Longitudinal patterns of cooperative interaction (reply).

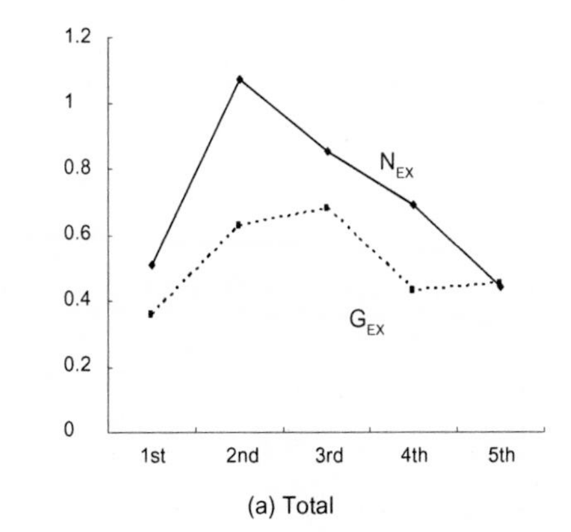

(a) Total

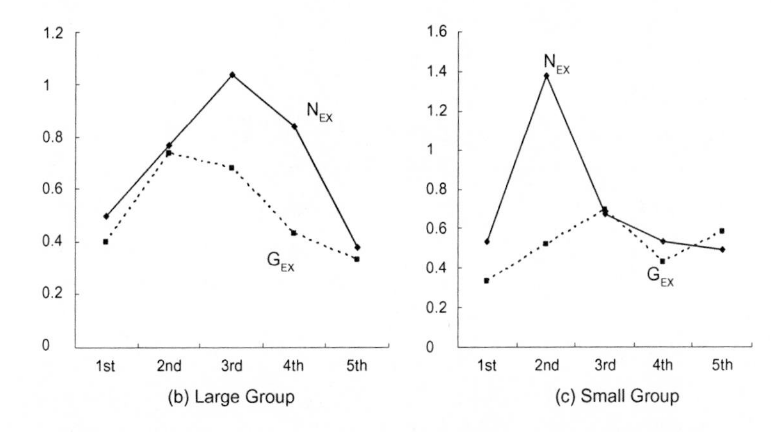

(b) Large Group

(c) Small Group

Appendix B

Questionnaires

I. Instruction for Participants

Welcome to the Online "Music" Forum

The Austin Association of Independent Musicians (AAIM) will sponsor a project aimed at promoting and encouraging people's conversations about music through the Internet. It is already a ubiquitous phenomenon—that people gather on the Web and exchange information, ideas, and opinions about various issues. Virtual communities and discussion groups are examples of such online social collectivities. This event will further this online phenomenon by focusing on online "music" conversations.

In this project, participants are supposed to communicate freely with others in an eight-person group setting during 5 consecutive days. The primary purpose of the group is to construct a useful online information pool through cooperative interactions among individuals. Individual members' contributions are crucial for the success of each group. As every member in the group actively contributes, the group's online information pool will become more and more rich and useful for the group members themselves.

After the 5-day period of online communication, AAIM will judge the best-performing groups based on the quantity of relevant information communicated. AAIM will give away a total of $1200, and will award each selected group $240 ($30 per person).

Important Points to be Noted:

1. Each group's performance will be evaluated on a daily basis. Individuals should post messages as frequently as possible to improve the daily performance of their group.
2. When entering the site, the first thing to do is to sign up with a preferred login name and password. The electronic board will be fully functional only when the account holder has logged in. Each individual will be identified by others only through his/her screenname.
3. Participants may address a virtually unlimited number of topics in the group. The only restriction is that they should be related to "music" as a product class in one way or another. Thus, possible topics may include participants' general relationships with various music genres, recommendations of favorite musicians and/or products, useful information/news or comments about musicians or recording companies, the new release of music albums, or simply questions and answers, among others.
4. Shortly after the 5-day period, participants will be asked to answer some questions regarding their online communication experiences. Only those who complete the online survey will be considered eligible for the awards.

If you have read and understood the instructions above, click on the button provided below to enter the place where you can meet other music fans. Please write down somewhere the address of this instruction page so you can access the virtual group site anytime during the period. If you need further assistance for participating in this project, please contact Dongyoung Sohn at dysohn@mail.utexas.edu.

II. Main Experiment Questionnaire

Virtual Community Study
IRB# 2004-09-0160

You have successfully communicated about "music" as a product category with other people for the past 5 days. This online survey is the last part of the virtual-community study you have participated in. It will take approximately 10 minutes for you to complete this questionnaire. All information you provide will be kept confidential and used for research purposes only. Data from the survey will only be presented in an aggregate form. To participate, you may go directly to the survey questionnaire online, where you will be asked a series of questions related to your Web usage experiences, opinions, and so forth. The risk in participating in this study is no greater than the risk associated with everyday living.

Your decision to participate or not will not affect your present or future relationship with The University of Texas at Austin. If you have any questions regarding this online survey, please feel free to contact Dongyoung Sohn, who is the Principal Investigator of this study, at dysohn@mail.utexas.edu.

Section 1. General Web Usage

Many people say that the World Wide Web (WWW) serves as a unique place where people efficiently communicate and exchange ideas, opinions, and information with one another. In particular, we may easily observe that there are many people on the Web trying to communicate product-related information, usage experiences, and opinions.

1. Have you ever participated in product-related discussions on the Web by expressing your ideas/opinions of any kind, or by giving out information you have?

   □ Yes
   □ No

2. If yes in Q1, please rate your degree of familiarity with "online product-related discussions."

   Not at all familiar □ □ □ □ □ □ □ *Extremely familiar*

3. If no in Q1, please indicate the reasons briefly below.

   ____

   ____

4. When purchasing a product, which one of the following do you think the most reliable information source for helping your decision making? Please assign a weight to each option by dividing 100 (total must be 100).

   [ ] Product Information from Companies
   [ ] Opinions of Other Consumers

Section 2. Your General Opinion

Notice: Please Read the Instructions Carefully

Please imagine a person who is randomly paired with you. You do not know the person, and will not knowingly meet in the future. We refer to the person as the "Other." Each point shown below has value. The more points you receive, the better for you, and the more points the "Other" receives, the better for him/her.

Here is an example:

	A	B	C
You get	500	500	550
Other gets	100	500	300

In this example, if you chose A you would receive 500 points and the other would receive 100 points; if you chose B, you would receive 500 points and the other 500; and if you chose C, you would receive 550 points and the other 300. So, you see that your choice influences both the number of points you receive and the number of points the other receives.

Before you begin making choices, please keep in mind that there are no right or wrong answers—choose the option that you, for whatever reason, prefer most. Also, remember that the points have value: the more of them you accumulate, the better for you. Likewise, from the "other's" point of view, the more he/she accumulates, the better for him/her.

	A	B	C
(1) You get	480	540	480
Other gets	80	280	480

	A	B	C
(2) You get	560	500	500
Other gets	300	500	100

	A	B	C
(3) You get	520	520	580
Other gets	520	120	320

	A	B	C
(4) You gets	500	560	490
Other gets	100	300	490

	A	B	C
(5) You get	560	500	490
Other gets	300	500	90

	A	B	C
(6) You get	500	500	570
Other gets	500	100	300

	A	B	C
(7) You get	510	560	510
Other gets	510	300	110

	A	B	C
(8) You get	550	500	500
Other gets	300	100	500

	A	B	C
(9) You get	480	490	540
Other gets	100	490	300

Section 3. Your General Perception

1. Below is a set of word pairs. Please click one box on each line that closely expresses your thoughts and feelings about "MUSIC" as a product category.

Important	□	□	□	□	□	□	□	*Unimportant*
Of no concern	□	□	□	□	□	□	□	*Of concern to me*
Exciting	□	□	□	□	□	□	□	*Unexciting*
Relevant	□	□	□	□	□	□	□	*Irrelevant*
Means nothing	□	□	□	□	□	□	□	*Means a lot to me*
Useless	□	□	□	□	□	□	□	*Useful*
Worthless	□	□	□	□	□	□	□	*Valuable*

Boring	□	□	□	□	□	□	□	*Interesting*
Appealing	□	□	□	□	□	□	□	*Unappealing*
Matters to me	□	□	□	□	□	□	□	*Doesn't matter*
Fascinating	□	□	□	□	□	□	□	*Mundane*
Not needed	□	□	□	□	□	□	□	*Needed*

2. Please indicate the extent to which you agree or disagree with the following statements.

   A. I know pretty much about music.

   Strongly disagree □ □ □ □ □ □ □ *Strongly agree*

   B. I do not feel very knowledgeable about music.

   Strongly disagree □ □ □ □ □ □ □ *Strongly agree*

   C. Among my circle of friends, I'm one of the "experts" on music.

   Strongly disagree □ □ □ □ □ □ □ *Strongly agree*

   D. Compared to most other people, I know less about music.

   Strongly disagree □ □ □ □ □ □ □ *Strongly agree*

   E. When it comes to music, I really don't know a lot.

   Strongly disagree □ □ □ □ □ □ □ *Strongly agree*

SECTION 4. YOUR COMMUNICATION EXPERIENCE

You have had conversations with other music fans via the Internet during the past 5 days. Based on your experience, please answer the following questions.

1. How strongly did you feel belongingness to the group?

   Not at all strong □ □ □ □ □ □ □ *Very strong*

2. How much commonality did you think you shared with the members of the group?

   Not much □ □ □ □ □ □ □ *Very much*

3. How close did you feel toward the members of the group?

   Not close □ □ □ □ □ □ □ *Very close*

4. How favorably did you feel toward the members of the group?

   Not favorable □ □ □ □ □ □ □ *Very favorable*

5. During the 5-day conversations, how much do you think you (and your group members) made contributions to the group's outcome? Please assign a weight to yourself and the other group members by dividing 100 (total must be 100).

   [ ] My contribution

   [ ] The others' contributions

6. How much satisfied are you with your group's overall performance?

   Not satisfied □ □ □ □ □ □ □ *Very satisfied*

7. What is your sex?

   □ Male
   □ Female

8. What is your age?

□ 20 or Under
□ 21–25
□ 26–30
□ 31–35
□ 36–40
□ Over 40

Abbreviations

ANCOVA	analysis of covariance
CMC	computer-mediated human-human interaction, or computer-mediated communication
CME	computer-mediated environment
df	degrees of freedom
eWOM	electronic word of mouth
F	Fisher's *F* ratio
HCI	human-computer (or message) interaction
HTML	hypertext markup language
MANCOVA	multivariate analysis of covariance
MANOVA	multivariate analysis of variance
MSE	mean square error
ns	nonsignificant
RSS	really simple syndication
SE	standard error
SS	sum of squares
SVG	social value orientation group variable
WOM	word of mouth
XML	extensible markup language

References

Alba, J., & Hutchinson, W. (1987, March). Dimensions of consumer expertise. *Journal of Consumer Research, 13*, 411–454.

Ariely, D. (2000, September). Controlling the information flow: Effects on consumers' decision-making and preferences. *Journal of Consumer Research, 27*, 233–248.

Arndt, J. (1967, August). Role of product-related conversations in the diffusion of a new product. *Journal of Marketing Research, 4*, 291–295.

Axelrod, R. (1984). *The evolution of cooperation*. New York: Basic Books.

Bagozzi, R. (1975, October). Marketing as exchange. *Journal of Marketing, 39*, 32–39.

Bagozzi, R., & Dholakia, U. (2002). Intentional social action in virtual communities. *Journal of Interactive Marketing, 16*(2), 2–21.

Balasubramanian, S., & Mahajan, V. (2001). The economic leverage of the virtual community. *International Journal of Electronic Commerce, 5*(3), 103–138.

Bandura, A. (1986). *Social foundations of thought and action*. Englewood Cliffs, NJ: Prentice-Hall.

Barabasi, A.-L., & Albert, R. (1999). Emergence of scaling random networks. *Science, 286*(15), 509–512.

Bargh, J., & McKenna, K. (2004). The Internet and social life. *Annual Review of Psychology, 55*, 573–590.

Ben-Porath, Y. (1980). The F-connection: Families, friends, firms, and the organization of exchange. *Population and Development Review, 6*(1), 1–30.

Bettman, J., Luce, M., & Payne, J. (1998, December). Constructive consumer choice processes. *Journal of Consumer Research, 25*, 187–217.

Bickart, B., & Schindler, R. (2001, summer). Internet forums as influential sources of consumer information. *Journal of Interactive Marketing, 15*, 31–40.

Bonacich, P. (1990, June). Communication dilemmas in social networks: An experimental study. *American Sociological Review, 55*, 448–459.

Brown, J., & Duguid, P. (2000). *The social life of information*. Boston: Harvard Business School Press.

Brown, J., & Reingen, P. (1987, December). Social ties and word-of-mouth referral behavior. *Journal of Consumer Research, 14*, 350–362.

Brucks, M. (1985, June). The effects of product class knowledge on information search behavior. *Journal of Consumer Research, 12*, 1–16.

Bucy, E. (2004). Interactivity in society: Locating an elusive concept. *The Information Society, 20*, 373–383.

Burt, R. (1987, May). Social contagion and innovation: Cohesion versus structural equivalence. *American Journal of Sociology, 92*, 1287–1335.

Chan, K., & Misra, S. (1990). Characteristics of opinion leader: A new dimension. *Journal of Advertising, 19*(3), 53–60.

Cho, C.-H. (1999). How advertising works on the WWW: Modified elaboration likelihood model. *Journal of Current Issues and Research in Advertising, 21*(1), 33–50.

Cho, C.-H., & Leckenby, J. (1999). Interactivity as a measure of advertising effectiveness: Antecedents and consequences of interactivity in Web advertising. In M. Roberts (Ed.), *Proceedings of the 1999 Conference of the American Academy of Advertising* (pp. 162–179).

Cho, C.-H., Lee, J.-G., & Tharp, M. (2001). Different forced-exposure levels to banner advertisements. *Journal of Advertising Research, 41*(4), 45–56.

Coleman, J. (1990). *Foundations of social theory*. Cambridge, MA: The Belknap Press of Harvard University Press.

Coleman, J. (1988, spring). Free riders and zealots: The role of social networks. *Sociological Theory, 6*, 52–57.

Constant, D., Sproull, L., & Kiesler, S. (1996). The kindness of strangers: The usefulness of electronic weak ties for technical advice. *Organization Science, 7*(2), 119–135.

Cook, K., & Whitmeyer, J. (1992). Two approaches to social structure: Exchange theory and network analysis. *Annual Review of Sociology, 18*, 109–127.

Coyle, J., & Thorson, E. (2001). The effects of progressive levels of interactivity and vividness in Web marketing sites. *Journal of Advertising, 30*(3), 65–77.

Dawes, R. (1980). Social dilemmas. *Annual Review of Psychology, 31*, 169–193.

de Cremer, D., & van Vugt, M. (1999). Social identification effects in social dilemmas: A transformation of motives. *European Journal of Social Psychology, 29*, 871–893.

Dellarocas, C. (2003). The digitization of word of mouth: Promise and challenges of online feedback mechanisms. *Management Science, 49*(10), 1407–1424.

Dichter, E. (1966, November/December). How word-of-mouth advertising works. *Harvard Business Review, 44*, 147–157.

Ekeh, P. (1974). *Social exchange theory: The two traditions*. Cambridge, MA: Harvard University Press.

Elliott, M., & Warfield, A. (1993). Do market mavens categorize brands differently? *Advances in Consumer Research, 20*, 202–208.

Emerson, R. (1976). Social exchange theory. *Annual Review of Sociology, 2*, 335–362.

Engel, J., Blackwell, R., & Kegerreis, R. (1969). How information is used to adopt an innovation. *Journal of Advertising Research, 9*(4), 3–8.

Engel, J., Kegerreis, R., & Blackwell, R. (1969, July). Word-of-mouth communication by the innovator. *Journal of Marketing, 33*, 15–19.

Feick, L., & Price, L. (1987, January). The market maven: A diffuser of marketplace information. *Journal of Marketing, 51*, 83–97.

Fiske, S., & Taylor, S. (1991). *Social cognition* (2nd ed.). New York: McGraw-Hill.

Flynn, L., & Goldsmith, R. (1999). A short, reliable measure of subjective knowledge. *Journal of Business Research, 46*(1), 57–66.

Foa, E., & Foa, U. (1980). Resource theory: Interpersonal behavior as exchange. In J. Spence, J. Thibaut, & R. Carson (Eds.), *Social exchange: Advances in theory and research* (pp. 77–94). Morristown, NJ: General Learning Press.

Fortin, D., & Dholakia, R. (2005). Interactivity and vividness effects on social presence and involvement with a Web-based advertisement. *Journal of Business Research, 58*(3), 387–396.

Freeman, L. (1979). Centrality in social networks: I. conceptual clarification. *Social Networks, 1*(215), 239.

Frenzen, J., & Nakamoto, K. (1993, December). Structure, cooperation, and the flow of market information. *Journal of Consumer Research, 20*, 360–375.

Fulk, J. (1993). Social construction of communication technology. *Academy of Management Journal, 36*(5), 921–950.

Gatignon, H., & Robertson, T. (1985, March). A propositional inventory for new diffusion research. *Journal of Consumer Research, 11*, 849–867.

Gatignon, H., & Robertson, T. (1998). Innovative decision processes. In T. Robertson & H. Kassarjian (Eds.), *Handbook of consumer behavior* (pp. 316–348). Upper Saddle River, NJ: Prentice-Hall.

Ghose, S., & Dou, W. (1998, March/April). Interactive functions and their impacts on the appeal of Internet presence site. *Journal of Advertising Research, 38*, 29–43.

Goffman, E. (1957). Alienation from interaction. *Human Relations, 10*, 47–59.

Gould, R. (1993). Collective action and network structure. *American Sociological Review, 58*(2), 182–196.

Granitz, N., & Ward, J. (1996). Virtual communities: A socio-cognitive analysis. *Advances in Consumer Research, 23*, 161–166.

Granovetter, M. (1973). The strength of weak ties. *American Journal of Sociology, 78*, 1360–1380.

Granovetter, M. (1974). *Getting a job: A study of contacts and careers*. Cambridge, MA: Harvard University Press.

Granovetter, M. (1978). Threshold model of collective behavior. *American Journal of Sociology, 83*(6), 1420–1443.

Granovetter, M. (1985, November). Economic action and social structure: The problem of embeddedness. *American Journal of Sociology, 91*, 481–510.

Ha, L., & James, E. (1998). Interactivity reexamined: A baseline analysis of early business Web sites. *Journal of Broadcasting and Electronic Media, 42*(4), 457–469.

Hagel, J., III, & Armstrong, A. (1997). *Net gain: Expanding markets through virtual communities.* Boston: Harvard Business School Press.

Hardin, R. (1982). *Collective action.* Baltimore: The Johns Hopkins University Press.

Heckathorn, D. (1996, April). The dynamics and dilemmas of collective action. *American Sociological Review, 61*, 250–277.

Hemetsberger, A. (2002). Fostering cooperation on the Internet: Social exchange processes in innovative virtual consumer communities. *Advances in Consumer Research, 29*, 354–356.

Henning-Thurau, T., Gwinner, K., Walsh, G., & Gremler, D. (2004, winter). Electronic word-of-mouth via consumer-opinion flatforms: What motivates consumers to articulate themselves on the Internet? *Journal of Interactive Marketing, 18*, 38–52.

Hernandez, M., Chapa, S., Minor, M., & Barranzuela, F. (2004). Hispanic attitudes toward advergames: A proposed model of their antecedents. *Journal of Interactive Advertising, 5*(1), 116–131.

Herr, P., Kardes, F., & Kim, J. (1991, March). Effects of word-of-mouth and product-attribute information on persuasion: An accessibility-diagnosticity perspective. *Journal of Consumer Research, 17*, 454–462.

Hoffman, D., & Novak, T. (1996, July). Marketing in hypermedia computer-mediated environments: Conceptual foundations. *Journal of Marketing, 60*, 50–68.

Katz, E., & Lazarsfeld, P. (1955). *Personal influence*. New York: Free Press.

Katz, M., & Shapiro, C. (1985). Network externalities, competition, and compatibility. *The American Economic Review, 75*(3), 424–440.

Kerr, N. (1989). Illusions of efficacy: The effects of group size on perceived efficacy in social dilemmas. *Journal of Experimental Social Psychology, 25*(4), 287–313.

Kollock, P. (1998). Social dilemmas: The anatomy of cooperation. *Annual Review of Sociology, 24*, 183–214.

Kollock, P. (1999). The economies of online cooperation. In P. Kollock & M. Smith (Eds.), *Communities in cyberspace* (pp. 220–239). New York: Routledge.

Komorita, S., & Parks, C. (1996). *Social dilemmas*. Boulder, CO: West View Press.

Kramer, R., McClintock, C., & Messick, D. (1986). Social values and cooperative response to a simulated resource conservation crisis. *Journal of Personality, 54*(3), 576–582.

Kraut, R., Patterson, M., Lundmark, V., Kiesler, S., Mukophadhyay, T., & Scherlis, W. (1998). Internet paradox: A social technology that reduces social involvement and psychological well-being? *American Psychologist, 53*(9), 1017–1031.

Kuk, G., & Yeung, F. (2002). Interactivity in e-commerce. *Quarterly Journal of Electronic Commerce, 3*(3), 223–234.

Lawler, E., & Yoon, J. (1996, February). Commitment in exchange relations: Test of a theory of relational cohesion. *American Sociological Review, 61*, 89–108.

Lawler, E., & Yoon, J. (1998, December). Network structure and emotion in exchange relations. *American Sociological Review, 63*, 871–894.

Leckenby, J., & Li, H. (2000). From the editors: Why we need the *Journal of Interactive Advertising. Journal of Interactive Advertising, 1*(1), 1–4.

Leibenstein, H. (1976). *Beyond economic man: A new foundation for microeconomics*. Cambridge, MA: Harvard University Press.

Levi-Strauss, C. (1969). *The elementary structures of kinship*. Boston: Beacon.

Li, H., & Bukovac, J. (1999). Cognitive impact of banner ad characteristics: An experimental study. *Journalism and Mass Communication Quarterly, 76*(2), 341–353.

Liebrand, W. (1984). The effect of social motives, communication and group size on behaviour in a N-person multi-stage mixed motive game. *European Journal of Social Psychology, 14*(3), 239–264.

Lin, N. (1971). *The study of human communications*. Indianapolis, IN: Bobbs-Merrill.

Lohtia, R., Donthu, N., & Hershberger, E. (2003). The impact of content and design elements on banner advertising click-through rates. *Journal of Advertising Research, 43*(4), 410–418.

Lombard, M., & Snyder-Duch, J. (2001). Interactive advertising and presence: A framework. *Journal of Interactive Advertising, 1*(2), 1–15.

Macias, W. (2003). A preliminary structural equation model of comprehension and persuasion of interactive advertising brand Web sites. *Journal of Interactive Advertising, 3*(2), 49–65.

Macy, M., & Willer, R. (2001). From factors to actors: Computational sociology and agent-based modeling. *Annual Review of Sociology, 28*, 143–166.

Mahajan, V., Muller, E., & Bass, F. (1990, January). New product diffusion models in marketing: A review and directions for research. *Journal of Marketing, 54*, 1–26.

Marwell, G., & Ames, R. (1979). Experiments on the provision of public goods. I. Resources, interest, group size, and the free-rider problem. *American Journal of Sociology, 84*(6), 1335–1360.

McMillan, S., & Hwang, J.-S. (2002). Measures of perceived interactivity: An exploration of the role of direction of communication, user control, and time in shaping perceptions of interactivity. *Journal of Advertising, 31*(3), 29–42.

McMillan, S., Hwang, J.-S., & Lee, G. (2003). Effects of structural and perceptual factors on attitudes toward the Website. *Journal of Advertising Research, 43*(4), 400–409.

Messick, D., & Brewer, M. (1983). Solving social dilemmas: A review. In L. Wheeler & P. Shaver (Eds.), *Review of personality and social psychology* (pp. 11–44). Beverly Hills, CA: Sage.

Messick, D., & McClintock, C. (1968). Motivation bases of choice in experimental games. *Journal of Experimental Social Psychology, 4*(1), 1–25.

Midgley, D. (1976). A simple mathematical theory of innovative behavior. *Journal of Consumer Research, 3*(1), 31–41.

Midgley, D., & Dowling, G. (1978, March). Innovativeness: The concept and its measurement. *Journal of Consumer Research, 4*, 229–242.

Miller, S. (1996). *Civilizing cyberspace: Policy, power, and the information superhighway*. New York: ACM press.

Molm, L. (1994). Dependence and risk: Transforming the structure of social exchange. *Social Psychology Quarterly, 57*(3), 163–176.

Morris, M., & Ogan, C. (1996, winter). The Internet as mass medium. *Journal of Communication, 46*, 39–50.

Muniz, A., & O'Guinn, T. (2001, March). Brand community. *Journal of Consumer Research, 27*, 412–432.

Myers, J., & Robertson, T. (1972, February). Dimensions of opinion leadership. *Journal of Marketing Research, 11*, 41–46.

Newhagen, J., Cordes, J., & Levy, M. (1995). Nightly@nbc.com: Audience scope and the perception of interactivity in viewer mail on the Internet. *Journal of Communication, 45*(3), 164–175.

Oliver, P., & Marwell, G., (1988, February). The paradox of group size in collective action: A theory of the critical mass II. *American Sociological Review, 53*, 1–8.

Olson, M. (1965). *The logic of collective action: Public goods and the theory of groups*. Cambridge, MA: Harvard University Press.

Ostrom, E. (1990). *Governing the commons: The evolution of institutions for collective action*. New York: Cambridge University Press.

Pavlou, P., & Stewart, D. (2000). Measuring the effects and effectiveness of interactive advertising: A research agenda. *Journal of Interactive Advertising, 1*(1), 5–25.

Peterson, R., & Merino, M. (2003, February). Consumer information search and the Internet. *Psychology and Marketing, 20*, 99–121.

Pieters, R., & Robben, H. (1998). Beyond the horse's mouth: Exploring acquisition and exchange utility in gift evaluation. *Advances in Consumer Research, 25*, 163–169.

Piliavin, J., & Charng, H.-W. (1990). Altruism: A review of recent theory and research. *Annual Review of Psychology, 16*, 27–65.

Poole, M., & DeSanctis, G. (1990). Understanding the use of group decision support systems: The theory of adaptive structuration. In J. Fulk & C. Steinfield (Eds.), *Organizations and communication technology* (pp. 173–193). Newbury Park, CA: Sage.

Rafaeli, S. (1988). Interactivity: From new media to communication. In R. Hawkins & J. Wiemann (Eds.), *Advancing communication science: Merging mass and interpersonal processes* (pp. 110–134). Newbury Park, CA: Sage.

Reingen, P., & Kernan, J. (1986, November). Analysis of referral network in marketing: methods and illustration. *Journal of Marketing Research, 23*, 370–378.

Rheingold, H. (1993). *The virtual community: Homesteading on the electronic frontier*. New York: Addison-Wesley.

Richins, M. (1983, winter). Negative word-of-mouth by dissatisfied consumers: A pilot study. *Journal of Marketing, 47*, 68–78.

Richins, M., & Root-Shaffer, T. (1988). The role of involvement and opinion leadership in consumer word-of-mouth: An implicit model made explicit. *Advances in Consumer Research, 15*, 32–36.

Robben, H., & Verhallen, T. (1994). Behavioral costs as determinants of cost perception and preference formation for gifts to receive and gifts to give. *Journal of Economic Psychology, 15*, 333–350.

Robertson, T., & Myers, J. (1969, May). Personality correlates of opinion leadership and innovative buying behavior. *Journal of Marketing Research, 6*, 164–168.

Robinson, J. (1976). Interpersonal influence in election campaigns: Two-step flow hypothessis. *Public Opinion Quarterly, 40*(3), 304–319.

Rodgers, S., & Thorson, E. (2000). The interactive advertising model: How users perceive and process online ads. *Journal of Interactive Advertising, 1*(1), 26–50.

Rogers, E. (1976, March). New product adoption and diffusion. *Journal of Consumer Research, 2*, 290–301.

Rogers, E. (2003). *Diffusion of innovations* (5th ed.). New York: Free Press.

Rogers, E., & Bhowmik, D. (1970). Homophily-heterophily: Relational concepts for communication research. *Public Opinion Quarterly, 34*(4), 523–538.

Rosen, E. (2000). *The anatomy of buzz: How to create word of mouth marketing*. New York: Doubleday.

Rosen, J., & Haaga, D. (1998). Facilitating cooperation in a social dilemma: A persuasion approach. *The Journal of Psychology, 132*(2), 143–153.

Rubin, A. (1994). Media uses and effects: A uses-and-gratifications perspective. In J. Bryant & D. Zillmann (Eds.), *Media effects* (pp. 417–436). Hillsdale, NJ: Lawrence Erlbaum Associates, Inc.

Rust, R., & Lemon, K. (2001). E-service and the consumer. *International Journal of Electronic Commerce, 5*(3), 85–101.

Sahlins, M. (1974). *Stone age economics*. Chicago: Aldine Atherton.

Schelling, T. (1978). *Micromotives and macrobehavior*. New York: W. W. Norton & Company.

Schlosser, A. (2003, September). Experiencing products in the virtual world: The role of goal and imagery in influencing attitudes versus purchase intentions. *Journal of Consumer Research, 30*, 184–198.

Silver, S., Cohen, B., & Crutchfield, J. (1994). Status differentiation and information exchange in face-to-face and computer-mediated idea generation. *Social Psychology Quarterly, 57*(2), 108–123.

Simmel, G. (1950). The triad. In K. Wolff (Ed.), *The sociology of Georg Simmel* (pp. 145–169). New York: Free Press.

Skvoretz, J. (1985). Random & biased networks: Simulations and approximations. *Social Networks, 7*, 225–261.

Skvoretz, J., & Lovaglia, M. (1995). Who exchanges with whom: Structural determinants of exchange frequency in negotiated exchange networks. *Social Psychology Quarterly, 58*(3), 163–177.

Skvoretz, J., & Willer, D. (1991). Power and exchange networks: Setting and structural variables. *Social Psychology Quarterly, 54*(3), 224–238.

Sproull, L., & Kiesler, S. (1986). Reducing social context cues: Electronic mail in organizational communication. *Management Science, 32*(11), 1492–1512.

Steuer, J. (1992, winter). Defining virtual reality: Dimensions determining telepresence. *Journal of Communication, 42*, 73–93.

Stewart, D., & Pavlou, P. (2002). From consumer response to active consumer: Measuring the effectiveness of interactive media. *Journal of the Academy of Marketing Science, 30*(4), 376–396.

Stromer-Galley, J. (2004). Interactivity-as-product and interactivity-as-process. *The Information Society, 20*, 391–394.

Sundar, S. (2004). Theorizing interactivity's effects. *The Information Society, 20*, 385–389.

Sussman, S., & Sproull, L. (1999). Straight talk: Delivering bad news through electronic communication. *Information Systems Research, 10*(2), 150–197.

Tajfel, H. (Ed.). (1978). *Differentiation between social groups: Studies in the psychology of intergroup relations*. London: Academic Press.

Takahashi, N. (2000). The emergence of generalized exchange. *American Journal of Sociology, 105*(4), 1105–1134.

Utz, S. (2004). Self-activation is a two-edged sword: The effects of I primes on cooperation. *Journal of Experimental Social Psychology, 40*(6), 769–776.

Valente, T. (1995). *Network models of the diffusion of innovations*. Cresskill, NJ: Hampton Press.

van Lange, P., de Bruin, E., Otten, W., & Joireman, J. (1997, October). Development of prosocial, individualistic, and competitive orientations: Theory and preliminary evidence. *Journal of Personality and Social Psychology, 73*, 733–746.

van Vugt, M. (1997). Concerns about the privatization of public goods: A social dilemma analysis. *Social Psychology Quarterly, 60*(4), 355–367.

Verhallen, T., & Pieters, R. (1984). Attitude theory and behavioral costs. *Journal of Economic Psychology, 5*, 223–249.

von Neumann, J., & Morgenstern, O. (1947). *Theory of games and economic behavior*. Princeton, NJ: Princeton University Press.

Walther, J. (1996). Computer-mediated communication: Impersonal, interpersonal, & hyperpersonal interaction. *Communication Research, 23*(1), 3–43.

Wasserman, S., & Faust, K. (1994). *Social network analysis: Methods and applications*. Cambridge, U.K.: Cambridge University Press.

Wellman, B. (2001). Computer networks as social networks. *Science, 293*(14), 2031–2034.

Wellman, B., & Hampton, K. (1999). Living networked on and offline. *Contemporary Sociology, 28*(6), 648–654.

Wellman, B., Salaff, J., Dimitrova, D., Garton, L., Gulia, M., & Haythornthwaite, C. (1996). Computer networks as social networks: Collaborative work, telework, and virtual community. *Annual Review of Sociology, 22*, 213–238.

Wiedmann, K.-P., Walsh, G., & Mitchell, V.-W. (2001). The mannmaven: An agent for diffusing market information. *Journal of Marketing Communications, 7*, 195–212.

Williams, T., & Slama, M. (1995). Market mavens' purchase decision evaluative criteria: Implications for brand and store promotion efforts. *Journal of Consumer Marketing, 12*(3), 4–21.

Wit, A., & Wilke, H. (1992). The effects of social categorization on cooperation in three types of social dilemmas. *Journal of Economic Psychology, 13*, 135–151.

Wu, G. (2000). *The role of perceived interactivity in interactive ad processing.* Unpublished doctoral dissertation, The University of Texas at Austin.

Yamagishi, T. (1988). Seriousness of social dilemmas and the provision of a sanctioning system. *Social Psychology Quarterly, 51*(1), 32–42.

Yamagishi, T., & Cook, K. (1993). Generalized exchange and social dilemmas. *Social Psychology Quarterly, 56*(4), 235–248.

Yamagishi, T., & Kiyonari, T. (2000). The group as the container of generalized reciprocity. *Social Psychology Quarterly, 63*(2), 116–132.

Zaichkowsky, J. (1985, December). Measuring the involvement construct. *Journal of Consumer Research, 12*, 341–352.

Index

Printed in the United Kingdom
by Lightning Source UK Ltd.
136359UK00001B/92/P